The Politics of GM Crops in India

‖‖‖‖‖‖‖‖‖‖‖‖‖‖‖‖‖‖‖‖‖‖‖‖
I0131779

This book discusses the conflicting discourse around GM crops in India. It brings together concerns related to food production, farming, environment, health, ownership and policymaking on the use of genetically modified crops in India.

The volume analyses apprehensions around GM technology from the perspective of the various stakeholders involved in the debate. Through field surveys and interviews with scientists, economists, environmentalists, civil society activists and cotton-growing farmers from the states of Telangana and Maharashtra, it highlights the vulnerabilities and questions related to the short-term and long-term impacts of using GM technology on farmers, food production, health, the agricultural economy and the environment. The book proposes ways to use GM technology, which takes stock of economic and farming limitations and accordingly brings in reforms and policies to reconcile the conflicting arguments of stakeholders.

This volume will be of great interest to researchers and students of development studies, political science, sociology, agricultural studies and sciences and biotechnology. It will also be useful for policymakers, think tanks and NGOs working with farmers or agriculture collectives on policy issues.

Asheesh Navneet is Assistant Professor of Political Science at the Tata Institute of Social Sciences, Tuljapur Campus, Maharashtra.

The Politics of GM Crops in India

Public Policy Discourse

Asheesh Navneet

R Routledge
Taylor & Francis Group

LONDON AND NEW YORK

First published 2021
by Routledge
2 Park Square, Milton Park, Abingdon, Oxon OX14 4RN

and by Routledge
52 Vanderbilt Avenue, New York, NY 10017

Routledge is an imprint of the Taylor & Francis Group, an informa business

© 2021 Asheesh Navneet

The right of Asheesh Navneet to be identified as authors of this work has been asserted by them in accordance with sections 77 and 78 of the Copyright, Designs and Patents Act 1988.

All rights reserved. No part of this book may be reprinted or reproduced or utilised in any form or by any electronic, mechanical, or other means, now known or hereafter invented, including photocopying and recording, or in any information storage or retrieval system, without permission in writing from the publishers.

Trademark notice: Product or corporate names may be trademarks or registered trademarks, and are used only for identification and explanation without intent to infringe.

British Library Cataloguing-in-Publication Data
A catalogue record for this book is available from the British Library

Library of Congress Cataloging-in-Publication Data
A catalog record for this book has been requested

ISBN: 978-0-367-35042-0 (hbk)
ISBN: 978-0-429-32943-2 (ebk)

Typeset in Sabon
by Apex CoVantage, LLC

Contents

Figures

Tables

Abbreviations

ABF	Alliance for Better Foods
ABLE	Association of Biotechnology-led Enterprises
ABLE-AG	Association of Biotechnology-led Enterprises-Agriculture Group
ACF	advocacy coalition framework
AFFB	Agriculture, Forest and Fisheries Branch
AIBA	All India Biotechnology Association
APAARI	Asia-Pacific Association of Agricultural Research Institutions
APCoAB	Asia-Pacific Consortium on Agricultural Biotechnology
APVVU	Andhra Pradesh Vyavasaya Vruthidarula
ASHA	Alliance for Sustainable and Holistic Agriculture
BJP	Bhartiya Janata Party
BKS	Bhartiya Kisan Sangh
BKS	Bharat Krishak Samaj
BRAI	Biotechnology Regulatory Authority of India
Bt	*Bacillus thuringiensis*
CA	Competent Authority
CAC	Codex Alimentarius Commission
CBD	Convention on Biological Diversity
CCFS	Central Committee for Food Standards
CCMB	Centre for Cellular and Molecular Biology
CESCR	Committee on Economic, Social and Cultural Rights
CIFA	Consortium of Indian Farmers Association
CIPRA	Centre for Indian Political Research and Analysis
CPB	Cartagena Protocol on Biosafety
CPI-M	Communist Party of India-Marxist
CREATE	Consumer Research, Education, Action, Training and Empowerment
CRO	Contract research organisation
CSA	Centre for Sustainable Agriculture
CSD	Council for Social Development
CSIR	Council of Scientific and Industrial Research
DBT	Department of Biotechnology

DLC	District-level committee
DNA	Deoxyribonucleic acid
EC	European Community
EPA	Environmental Protection Act
EPA	Environmental Protection Agency
EU	European Union
FAO	Food and Agricultural Organisation
FEDCOT	Federation of Consumer Organisations of Tamil Nadu
FFA	Federation of Farmers Association
FFDCA	Federal Food, Drug and Cosmetic Act
GCA	Gross cropped area
GDDP	Gross district domestic product
GEAC	Genetic Engineering Approval/Appraisal Committee
GM	Genetically modified
GMOs	Genetically modified organisms
GOI	Government of India
HAHB	Human and Animal Health Branch
HT	Herbicide tolerant
IAASTD	International Assessment of Agricultural Knowledge, Science and Technology for Development
IBSC	Institutional Biosafety Committee
ICAR	Indian Council of Agricultural Research
ICGEB	International Centre for Genetic Engineering and Biotechnology
ICMR	Indian Council of Medical Research
IEAB	Industrial and Environmental Applications Branch
IGMORIS	Indian GMO Research Information System
IISC	Indian Institute of Science
IPR	Intellectual property rights
LDF	Left Democratic Front
LL rice	Liberty Link rice
MEC	Monitoring and Evaluation Committee
MHFW	Ministry of Health and Family Welfare
MMB	Mahyco-Monsanto Biotech. India Ltd.
MoA	Ministry of Agriculture
MoEF	Ministry of Environment and Forests
MoEFCC	Ministry of Environment, Forests and Climate Change
MoS&T	Ministry of Science and Technology
NA	Not applicable
NBPGR	National Bureau of Plant Genetic Resources
NBRA	National Biotechnology Regulatory Authority
NDA	National Democratic Alliance
NEPA	National Environmental Policy Act
NGO	Non-governmental organisation
NIAP	National Institute of Agricultural Economics and Policy Research

NIH	National Institutes of Health
NIPGR	National Institute of Plant Genome Research
NITI	National Institution for Transforming India
N-KLSP	Nagoya-Kuala Lumpur Supplementary Protocol
NOC	No Objection Certificate
NRC	National Research Council
OBC	Other Backward Class
OSTP	Office of Science and Technology Policy
PHC	Public Health Centre
PIL	Public Interest Litigation
PM	Prime minister
PMO	Prime minister's office
RCGM	Review Committee on Genetic Manipulation
RDAC	Recombinant DNA Advisory Committee
rDNA	Recombinant deoxyribonucleic acid
R&D	Research and development
RFSTE	Research Foundation for Science, Technology and Ecology
RSS	Rashtriya Swayamsevak Sangh
RTI	Right to information
SAUs	State agricultural universities
SBCC	State Biotechnology Coordination Committee
SC/ST	Scheduled Caste/Scheduled Tribe
SDDS-DES	Data Dissemination Standards-Directorate of Economics and Statistics
SJM	Swadeshi Jagaran Manch
SOPs	Standard operating procedures
SSKP	Shaswat Sheti Kriti Parishad
TEC	Technical Expert Committee
TOI	*Times of India*
TRIPS	Trade-Related Intellectual Property Rights
TV	Television
UK	United Kingdom
UN	United Nations
UPA	United Progressive Alliance
US	United States
USDA	United States Department of Agriculture
USFDA	United States Food and Drug Administration
WTO	World Trade Organization

Acknowledgements

This book is an extended work of my PhD thesis. I thank all those who supported and helped me in completing this book successfully. I particularly thank the ISEC and TISS society for providing me all the facilities needed for my research study. I thank the Indian Council of Social Science Research (ICSSR) for granting me the fellowship for three years to conduct this research. I thank the Malcolm and Elizabeth Adiseshiah Trust for granting me the fellowship for six months to complete my research.

I thank and extend my deep sense of gratitude to my PhD supervisor, Dr V. Anil Kumar, for his guidance, constant moral support and encouragement to complete the writing. He provided me with several important readings that helped me in understanding my research area better and in improving the thought process while critically analysing different bodies of literature. It was a pleasure to work under his guidance.

I thank him for forming a strong and healthy doctoral committee consisting of experts from economics, sociology, and political science backgrounds. These experts were Dr Parmod Kumar from agricultural economics, Dr Sobin George and Dr Anand Inbanathan from sociology and Dr V. Anil Kumar himself from the political science department. I am grateful to all of them for providing their precious time to attend all the DC meetings to discuss my PhD work critically.

I also extend my thanks to all the panellists of biannual seminars invited by my supervisor, starting with Dr Elumalai Kannan, Dr Padma Sarangapani, Dr D. Rajasekhar and Dr M.V. Nadkarni, for providing their valuable comments and suggestions, which helped me to improve my research gradually. I also thank Dr Ramesh of the Law School for consenting to be the special discussant of some of the ISEC biannual seminars where I presented my work. I also thank Dr A.V. Manjunatha and Dr P.G. Chengappa for showing interest in my work and inviting me to participate in one of the workshops related to my research. The workshop turned out to be a fruitful one, as it consisted of scientists and social scientists gathered to discuss the future of transgenic crops in India. I am very grateful to Dr Chengappa for personally introducing me to the experts working in some of the major government regulatory bodies of India.

I should not forget to thank my friends who helped and supported me during my field study that I undertook in Delhi, Hyderabad, Warangal and Yavatmal. First of all, I would like to thank my sister, Dr Sudha Shashi, for helping me to get accommodations in Delhi for my field study. She took good care of me during my one-month stay in Delhi.

I thank my friend, Dr Lalatendu Kesari Das, for sharing his room with me during my stay at Hyderabad. I thank all the academicians, scientists, social scientists and activists who took time out from their busy schedule to answer my questions.

I thank my friend, Omkar Patella, for accompanying me to the Warangal district of Telangana and introducing me to his friends there. Before leaving, he ensured that I got good accommodations at Kakatya University with his friend, Fanikant, who himself was a PhD scholar there. I befriended him and shared his room during my stay at Warangal. I extend my gratitude to Fanikant and his friend Sravan and others for helping me to interact with farmers in Telugu, which I do not understand. Sravan and his friends helped me by interpreting English to Telugu and Telugu to English. I thank all the farmers in Warangal who agreed to give interviews. I also thank the veterinary doctors and the livestock assistant in Warangal and Hyderabad, who agreed to talk to me on the issue of animal health.

I thank the Dilasa organisation team and its founder, Mr Madhukar Dhas, for providing me accommodations in the organisation's campus during my stay at the Yavatmal district of Maharashtra. The organisation provided me an assistant named Umesh, who showed me villages and helped me to interact with farmers. I am much obliged to him for that. I also thank all the farmers in Yavatmal who were good enough to allow me to interview them.

I thank the anonymous referees for reading the entire manuscript and giving insightful and valuable comments for the overall improvement of the book.

Last but certainly not least, I thank my family, in particular, my mother Ms B.J. Rani, my father Mr O.P. Verma and my sister Dr Sudha Shashi, for all that they have done for me all these years. I know that this acknowledgement is an insignificant act as their deep affection, love, support, and patience for me is something which I can never express fully in words.

1 Introduction
The historical background

Out of the nation's 1 billion people, two-thirds still earn their livelihood from farming and more than 75 per cent of Indian farmers are disadvantaged with regard to technology and land-holdings. Therefore, the utmost need for Indian farmers in the current scenario is to get access to sophisticated modern technology that would help them increase agricultural production. Biotechnology in this regard is seen as "a new sunrise industry, one that could be at the forefront of an economic transformation, replicating the success of the information technology (IT) sector" (Scoones, 2006).

Biotechnology and genetically modified (GM) crops, another recent innovation, are not the same. Biotechnology encompasses all technologies based on living forms, i.e. all biological technologies (Kuruganti & Ramanjaneyulu, 2008), whereas the GM crop is only one among several technologies under the broad umbrella of biotechnology. This technology was first developed by US scientists. It is new and different in the sense that for the first time, genes of a particular seed plant are modified in order to enable it to withstand adverse weather conditions.

Gene modification technology can also be referred to as recombinant DNA (rDNA) technology or gene-splicing technology. GM crops are different from their conventional counterparts in the sense that they contain a specific gene or set of genes that have been artificially inserted instead of the plant or crop acquiring them through pollination. Therefore, with the adoption of GM technology, scientists can now transfer the gene of a particular quality of one organism into another. The technology was first developed in the 1970s in the United States (Nestmann, Copeland, & Hlywka, 2002).

However, when these new GM crops were introduced into the European markets, a whole range of problematic issues concerned with health and environment arose. In India, apart from health and environment as problematic issues, others related to ownership rights have also been raised. Unlike in the past, farmers cannot generate GM crops by themselves, and they have become increasingly dependent on corporate sectors to get new GM seeds for the next season. Therefore, there lies a complex reality behind easy narratives about science and technology policy. The aim of this book is to understand critically the motive (political, social and economic) of choosing a particular technology and to understand and explicate the reasons for contestations and debates around them.

According to biotechnology scientists, genetically modified organisms (GMOs) have great potential to uplift the Indian economy with their various applications for human and animal health care systems, agricultural production, industrial products and environmental management. However, it has also been realised that "there could be unintended hazards and risks from the use of GMOs and products thereof, if the new technology is not assessed before use" (Ghosh & Ramaniah, 2002). In this book the analysis and focus will be more on its applications in agriculture. Various studies on the utility of this technology in the field of agriculture show several complications about the nature of this technology and its relationship with different sectors and societies.

Potentialities and scopes

Several GM crops are now grown around the world. Common among them are soybeans, corn, cotton and canola. Beside these, other GM crops that are being proposed are tomatoes, potatoes, papayas, chicory, melons, rice, squash, sugar beets, wheat, mustard and brinjal (Raman, 2017). However, in India only GM cotton has been approved for cultivation in the agricultural farmlands to date. The other GM crops are still in the process of getting approval from the government's agencies, like the Review Committee on Genetic Manipulation (RCGM) and Genetic Engineering Appraisal Committee (GEAC), and they are going through several tests in the form of trials and experiments.

One of the most common purposes of GM crops is to confer tolerance to herbicides that are sprayed on the farm lands to control weeds (Wu & Butz, 2004). The herbicide-tolerant crops include transgenes providing tolerance to the herbicide glyphosate or glufosinate ammonium. These herbicides have the ability to kill nearly all kinds of plants except those that have the tolerance gene. Further, the genetic modification in crops can also enable particular crops to protect themselves against harmful insects and pests. The uses of GM crops can also help the farmers in reducing the need for chemical pesticides that are sprayed to protect the crops against pests and in this way improve farmers' health by cutting down exposure to harmful and poisonous chemicals.

The soil bacterium called *Bacillus thuringiensis* is a new discovery by biotechnology scientists. By transplanting the gene of this particular bacterium into some agricultural plants like cotton and corn, scientists have found that these Bt plants can now produce crystal proteins that are toxic to certain insects but are generally harmless to vertebrates and non-lepidopteran insects (Wu & Butz, 2004; Raman, 2017). In this way, the genetic insertion of a bacterial gene into the genome of a plant enables it to produce its own pesticide.

Biotechnology scientists and the proponents of GM crop agriculture argue that the future of GM crops in agriculture is bright. This is because in the future scientists would certainly find ways to introduce GM crop agriculture into conditions that have remained adverse and unsuited for agriculture until now. For example, scientists have developed GM tomatoes that could be grown in salty soil. This could be useful, since other crops could not be grown due to the saline nature of the soil (Wu & Butz, 2004; Raman, 2017). There are other examples where scientists have enabled some rice varieties, through gene transfer technology, to stand for longer periods in flooded regions like Bengal in India and Nihe village in China. By introducing a gene called *FRi3A* into a high-yielding variety of rice, scientists have enabled the rice plants to not only survive longer in flooded water but also retain their productivity despite external influences like floods (Ronald & Adamchak, 2008; Raman, 2017). They are working to develop GM crops that could be grown also in drought areas.

Wu and Butz write in the "Future of Genetically Modified Crops" that with the help of GM technology some food crops like rice can be genetically modified to produce micronutrients like vitamin A, which are vital for human diets. Scientists have already developed a crop called 'Golden Rice' that can produce 'beta-carotene', a pigment deemed to be the precursor of vitamin A. This ability of GM technology clearly demonstrates its potential to solve problems like hunger and malnourishment, which are prevalent in developing countries like India. Further, food crops can also be genetically engineered to produce edible vaccines against infectious diseases that would then be readily made available to the children prone to these dreadful diseases. This would not only cure their diseases but would also provide them with sufficient nourishment. However, these are simply potentialities of GM crop technology that have yet to be realised outside the laboratories of scientists.

Major concerns

Scholars who have done some critical studies in the area are Vandana Shiva, Suman Sahai, Glenn Davis Stone, Andrew Flachs, Kavitha Kuruganti, Lianchawii, R.V. Anuradha and many more. According to Shiva's article "Globalisation and Threat to Seed Security", the Indian seed industry is rapidly coming under the control of corporate companies like Monsanto. In fact, Monsanto has entered into an exclusive agreement with the Maharashtra hybrid seed company Mahyco, and the two have formed a joint venture to introduce genetically modified Bt cotton in India.

Shiva et al. have argued that if large areas are sown with limited and uniform varieties of seeds, it would lead to increased risk for the farmers, as these varieties may become vulnerable to diseases and attacks by pests (Shiva, Emani, & Jafri, 1999). According to them, the new trend that is being set by the transnational companies through the use of genetically engineered plants is to create a niche for the broad international market for a single product. In addition, patent protection and intellectual property rights (IPR) imposed through the Trade-Related Intellectual Property Rights (TRIPS) agreement of World Trade Organization (WTO) will inhibit farmers from re-using, sharing and storing seeds. This increases the prospect that only a few varieties will be able to dominate the seed market.

Further against the IPR, Shiva et al. mentions that although ownership and property claims have now been made on living resources, the prior custody of these resources by farmers was never considered while setting up the patent rights. Shiva et al. stresses the fact that it is the intervention of technology that determines the claim for the exclusive use of plant seeds. In other words, possession of some form of a particular technology becomes the reason for ownership claims by companies like Monsanto and others.[1]

Suman Sahai (1999) in her article "What Is Bt and What Is Terminator?" draws our attention to a new genetic tool called 'terminator', which can design seeds with in-built sterility. The 'terminator' technology is also based on a genetic construct. Here two gene systems have been brought into play, only to abort the development of the embryo so as to cut down the germination processes of seeds.[2] According to Sahai, these self-destructing seeds are actually hybrids produced by hybridising two transgenics, each containing one of the two gene systems. Sahai states a very interesting point, which is that even when the two gene systems are brought together in the hybrid seed, they are viable and can germinate.

> In order to control the induction of sterility, a chemical switch has been built in. The switch can be activated by soaking the seeds in tetracycline. Once the tetracycline soaks into the seed tissue, it switches on one of the gene systems which set in motion the chemical process which will abort embryo. So in practice, the seed company can produce as much of the seed as they want and just before selling it to the farmer, they can treat the seeds with tetracycline to switch on the sterility inducing gene system.
>
> (Sahai, 1999)

Sahai further argues that when the farmer buys seeds from the company, he can grow one crop from it, but the seed produced in his crop when it matures is sterile and cannot germinate. Thus the farmer will not be in a position to save the seeds from his crop for the next sowing and will be forced to return to the seed company for new seeds (Sahai, 1999). Therefore this technology can help seed companies establish total control over the production and sale of seeds. As a result, the new terminator technology is miles ahead of the patent system in establishing a monopoly of multinational seed companies on the seed markets of the developing world.[3] According to Sahai, the multinational companies have an avowed goal to capture the vast potential of the seed business in India, China and other Southeast Asian countries.

Already with the adoption of monoculture practice by farmers, agriculture has become increasingly threatened by several unknown plant diseases, insect pests and weeds. Scholars like Shiva and Kuruganti argue that in the case of Bt cotton farming, resistance has started evolving among insects, weeds and pathogens, and therefore Bt cotton plants are not any safer against their harmful pests. Farmers now would have to incur an extra expenditure in buying chemical pesticides. Therefore, in a way, corporate companies are creating conditions to compel farmers to get completely dependent on them, which cannot be deemed ethical or moral.

Besides all these complexities and risks involved in GM crops, a study conducted by Kavitha Kuruganti in her book *Genetic Engineering in Indian Agriculture* shows some of the significant evidence that GM crops can also cause harm to living beings if consumed without proper testing (Kuruganti & Ramanjaneyulu, 2008). For instance, Kuruganti has discussed in her book independent research conducted by Dr Arpad Puzstai in the Rowett Research Institute in Scotland, UK. He conducted a multicentric study on rats fed with GM potatoes for three years. He, along with his research team, found some very unexpected and worrying changes in the size and weight of the body organs of rats fed GM potatoes. According to him, the liver and heart of rats fed with GM potatoes started getting smaller, and so did the brain. There were also indications that the rats' immune systems were weakening. Because he reported this information on a TV show, Dr Puzstai was fired from his job and discredited by the scientific community (Kuruganti & Ramanjaneyulu, 2008).

Sahai asks a very relevant question to the proponents of GM crops in India: what would be the use for GM herbicide-tolerant crops in India? According to her, it would displace labour that today does the weeding in fields. This would mean loss of income for farm labour, especially women, and also it would mean loss of fodder and nutritive leafy greens. According to her, what is a weed in a rice or wheat crop is valued as vegetables by rural families. For example, the highly nutritious chaulai and bathua Saag are highly valued by poor people in Indian villages. Some of the weeds also go for feeding the livestock. Therefore, according to Sahai, weeds support rural livelihoods, and so it would be nothing but foolishness to introduce GM herbicide-tolerant crops in these Indian conditions.[4]

Complexities involved in the use of technology

Several political and research questions about GM crops have been raised and have thus given rise to diverse positions among scholars and scientists. In India there are scholars like Vandana Shiva, who are highly critical of the uses of science and technology. This is because they believe that science and technology support patriarchy in the society. According to Shiva, regeneration is the basis for the sustainability of life. However, men created a patriarchy and started getting themselves away from nature by introducing science and technology. The real form of creativity, which was based on the process of 'regeneration', was denied the term 'creativity', and it was associated with the new form of technology of 'non-regeneration'. This non-ecological view of nature and culture formed the basis of patriarchal perceptions.[5]

By focusing on the earth, seeds and women's bodies as sites of regeneration, Shiva attempts to study how new biotechnologies in the form of GM crops are re-establishing the same old patriarchal divisions. The land, the forests, the rivers, the oceans and the atmosphere have all been colonised, eroded and polluted. According to her, private capital is now looking for new colonies, and these new colonies are nothing but the interior spaces of the bodies of women, plants and animals, which are being encroached on through the technology of genetic engineering only to control the regeneration process.[6]

Therefore, Shiva is a kind of scholar or intellectual who has expressed her viewpoint against science and technology. But the important question that comes to mind is: can we imagine human life without technology? If we look around, the human world is surrounded by technologies and scientific inventions. So are we really in a position to discard or ignore them? Considering the kind of impact that science and technology have on our lives, it is difficult to imagine our lives without them.

On this account, Sahai argues that since technology cannot be dissociated from human life, it is our responsibility to regulate it with honest and cautious effort instead of discarding it. According to her, technology must be regulated cautiously because scientists have acknowledged its safety concerns. She further states that regulation was sought not by the political leaders, not by the civil society, not by NGOs, but by the scientists. Therefore, we must take seriously the question of the safety of GM products and the fact that they need regulation. She further argues that in a country like India, GM technology, or any other technology, can certainly play a role, but to give credence to the technology beyond what the technology so far has shown is perhaps misguided.

For Sahai, GM seed technology is a very imprecise technology. This is because:

> We can neither guide the gene to where we want it, nor can we get the number of copies of the gene that we want. For instance, if I want to put in two genes in a genome of a cell, I can't. I shoot-in gene into a cell and then wait for something to happen. So there is a great randomness to this technology which is all right provided you take on that randomness on board and then work with the fact that it is not a precise technology and therefore you have to deal with the fact of safety testing prior to GM use.[7]

She further argues that a country like India has a policy of mandatory labelling of GM foods. Yet when permission was given by the Genetic Engineering

Appraisal Committee (GEAC)[8] for Bt-modified brinjal release, the country had not yet got a labelling infrastructure mechanism in place. So according to Sahai, we are in violation of our own rule. She further states that our government still does not have a proper liability clause and an adequate law that will grant compensation and redress if something goes wrong. According to her, neither science nor technology operates in vacuum. Just as science can do a lot of good, its application can also do a lot of harm. For instance, during the 1860s, an Austrian priest called Gregor Mendel laid the foundation of 'genetics' for understanding heredity, which has been of extreme importance in understanding human diseases. At the same time, through this understanding a medical technology called amniocentesis has been developed which has been used to determine sex and the killing of girl foetuses.

Therefore, there is a purity about science that we are all for, but there may not always be purity in its application when science gets turned into technology. Hence, the argument from the scientific community is that honest science should test whatever technology it invents until we are fairly confident that it is safe, and if it is not, then it should immediately be removed. Keeping this view in mind, we can now understand why 51 independent international scientists with expertise in genetic engineering and biosafety protocols have approved the Technical Expert Committee (TEC) panel's interim report until a final report is ready. "The interim report has called for a ten-year moratorium on open field trials of Bt food crops until adequate regulatory mechanisms and safety standards are put in place."[9]

Position of the Indian government on the issue of GM technology

When we closely observe the stand of government on the use of GM seed technology, we find that there is no unanimity, with different ministries expressing different opinions. The Ministry of Agriculture and Ministry of Science and Technology under the UPA government were trying their best to promote this technology for the growth of agriculture. On the other hand, the Ministry of Environment and Forests and Ministry of Health and Welfare were seeking more time for sufficient tests on the safety of environment and health before granting approval.

According to Aarti Gupta, in pursuance of 1989 biosafety rules, the authority to regulate GMOs is explicitly divided between the Ministry of Science and Technology and the Ministry of Environment and Forests (MoEF) (Gupta, 2000). Here comes the real conflict between the two ministries on the issue of granting approval to GM seed technology. The Department of Biotechnology (DBT) comes under the domain of the Ministry of Science and Technology, and one of the most important organs of the DBT is the Review Committee on Genetic Manipulation (RCGM).

Similarly, the Genetic Engineering Appraisal Committee (GEAC) falls under the ambit of the Ministry of Environment, Forests and Climate

Change (MoEFCC),[10] and the environment minister has the discretionary power to withhold the approval of GM crops. Therefore, even if any department which falls under the ambit of the Ministry of Science and Technology or Agriculture gives approval to GM crops, it cannot be enforced unless and until MoEFCC also gives its consent. Because of this, the conflict between these ministries is explicitly visible. The political tension and conflict among ministries can become clearer if we observe the progress during the UPA's term.

The previous Union Minister of MoEFCC, Jayanthi Natarajan, opposed the then Agriculture Minister Sharad Pawar's view in favour of GM technology and urged then Prime Minister Manmohan Singh to let her ministry take an independent view on GMOs. In a letter to Dr Singh, Natarajan said the Agriculture Ministry's mandate is to promote GM crops, whereas her ministry's role is to regulate their use.[11] Therefore, an independent view over the whole issue is needed. She also urged the PM to keep the field trials of GM crops in abeyance until parliament passes a bill to establish the Biotechnology Regulatory Authority of India (BRAI).[12]

The Supreme Court, on the other hand, appointed a Technical Expert Committee (TEC) to look into the matter, which submitted its report recommending a moratorium on field trials of GM crops until a regulatory mechanism is created (TEC, 2013). The government's view so far has been rooted in the joint position taken by the two ministries. However, Ms Natarajan's plea to put on hold the clearances of GM crop field trials explicitly laid out her opposition to the views of Mr Pawar[13] and the GM crop developer industry, which lobbied for clearances.

Finally Natarajan had to resign from the ministry. This gave rise to the speculation that corporate houses and industry and other ministries were not happy with her decision to put the field trials' clearances on hold. Just after her resignation, then Petroleum Minister Mr Veerappa Moily took charge of the Environment Ministry and soon approved the GM crop field trials despite the fact that the Supreme Court is still hearing the case.[14] The argument that Mr Moily has given in support of this move is that the apex court has never explicitly ordered any stay against clearing the field trials. The progress so far has only cleared the first hurdle in the way of GM crops. Other hitches remain, and a major one concerns the position of the state governments. Under the Indian Constitution, agriculture falls in the state list, and in 2011, GEAC, which is a prime body of the MoEFCC, mandated that field trials of any GM crop cleared by it would also have to be permitted by the state governments. It is interesting to know that according to a survey conducted by *Business Standard*, a majority of Indian states are against the move.[15] Therefore, consensus has still eluded various bodies of government on the policymaking process.

India is not the first to adopt GM technology. It was in the United States where this technology was first brought into existence and adopted. The GM crops have helped farmers in reducing their total expenditure on chemical pesticides by making the plants resistant to the attacks of insects like boll-worms. Despite this advantage, planting of GM crops has not spread signifi-cantly (Paarlberg, 2001).

Food politics in the EU and United States

Food is not just a commodity or nourishment. Its meaning and significance have to be understood at a much broader level (Brom, 2004). It is a basic need that is not met in several parts of the world. The Food and Agricul-tural Organization (FAO) has estimated that presently nearly 800 million people are undernourished (Brom, 2004). Therefore, according to Brom, food trade is a matter of great significance in dealing with the problems and hurdles related to "Global Justice and the internationally recognised rights". In many regions, food is also recognised as part of a cultural or religious identity. This can become more apparent if one makes a compara-tive analysis of the ways and forms of food practices in Jewish, Islamic and Hindu religions. According to Brom, "the production and consumption of food is one of the most expressive symbolisations of human interaction with nature, other living beings and the universe" (Brom, 2004). The UN Committee on Economic, Social and Cultural Rights (CESCR) asserted that "food should not only be safe i.e. free from adverse substances, but also be acceptable within a given culture" (Brom, 2004). Therefore, to create an adequate balance between cultural acceptability and safety issues, the problems related to collision between food treated as a commercial product by the corporate sector and food accepted as part of religious and cultural identity by ethnic groups has to be resolved. This cultural-religious aspect and safety issues related to food have gradually become the main bone of political and economic contention between the EU and the United States. It has particularly led to two of the transatlantic trade conflicts between them. These relate to the:

1   use of artificial growth hormones in beef production and
2   use of biotechnology in food crops.

There is a tussle going on between the United States and the EU regarding the use of GM crops. The two sides represent the two largest economies in the world. According to Pollack and Shaffer (2009), both polities are part of the advanced industrialised world, both are democratic and both feature federal or quasi-federal political systems where power has to be shared between states and a central government. Despite such similari-ties, the United States and EU have adopted different approaches towards the regulation of biotechnology in agriculture. It would be interesting to analyse the historical background of their regulatory approaches in order to understand the reason for their adoption of a particular regula-tory stand.

Reason for the differences between US and EU regulatory stands

The conventional reason given for US–EU differences on GM regulation is that Europeans care more for the natural environment than Americans (Anderson & Jackson, 2006). However, Anderson and Jackson believe that this EU belief is unworthy for two reasons:

1   Consumers and environmentalists do not wield a great deal of political clout in relation to the interest of producers.
2   There is no concrete evidence to justify the concerns reflected in the precautionary stance taken by the EU.

Among the European communities, the worries over GM technology that have arisen are related to food safety concerns. The EU regulatory agencies suspect that GM-derived foods might be toxic or carcinogenic or allergic for animals and humans and might be nutritionally less adequate. The EU scientific community also worries that transgenes of GM crops, if consumed, might survive digestion and might alter the genome of the person or animals consuming them. However, Anderson and Jackson have shown in their article that according to the report of the Food Safety Department of the UN's WHO, GM foods currently in the international market have passed successfully the risk assessments and therefore are not at all harmful for human or animal health (Anderson & Jackson, 2006).

The US regulatory approach

The regulatory supervision of GM technology has been in place for a longer time in the United States than in most parts of the world (Nestmann, Copeland, & Hlywka, 2002). However, the global dominance of the United States in the field of agricultural biotechnology has occurred without any proper planning or strategic industrial-targeting programmes (Isaac, 2002).

In 1991, the President's Council on Competitiveness, which is an American non-profit organisation based in Washington, D.C., recommended two key principles of US biotechnology regulations for technological progress (Isaac, 2002). They are:

1   focus on product standards, rather than technology or process standards
2   A particular form of regulation under different regulatory departments.

Thus, according to Isaac, both these principles are key features of the scientific rationality approach encouraging technological progress. In his book Isaac further mentions an important event that took place on 30 November 1999, when the US-based Alliance for Better Foods[16] held a news conference in Seattle, Washington, to publicly support the agricultural biotechnology and the US regulatory system. The most notable event in the conference was that of a statement of support for agricultural biotechnology and the US regulatory approach, which was undersigned by 330 scientists (Isaac, 2002). The importance of these two examples, according to Isaac, lies in the fact that the US regulatory approach has been explicitly linked to the promotion of biotechnology by establishing stable and predictable regulatory rules that encourage technological progress.

According to Nestmann, Copeland, and Hlywka (2002), GM technology is more advanced and precise, and it has been proven to predict more accurately the desired plant crops through genetic manipulation. This kind of prediction is not possible in the conventional plant breeding methods of agriculture. Hence, the conventional method of agriculture is associated with lot of uncertainties.

The US regulatory framework

The regulation of food and environmental safety in the United States was a matter for state and local governments in the past (Pollack & Shaffer, 2009). They took the primary responsibility of inspecting all kinds of food for safety concerns during the nineteenth century. However, by the beginning of the twentieth century, interstate trade relations had grown in the United States. Therefore, in order to keep the trade relations effective and efficient, the United States needed to make the food safety regulations standard across the states so that the regulations can be accepted and followed by all states on equal terms (Pollack & Shaffer, 2009).

Keeping this view in mind, the US Congress adopted the first comprehensive federal food safety legislation, viz. the Food and Drugs Act and Federal Meat Inspection Act in 1906–1907 under the Interstate Commerce Clause of the Constitution (Pollack & Shaffer, 2009). Further, these acts and the subsequent congressional legislation led to the establishment of a federal system that would look after the regulatory standards.

Federal environmental regulation came much later, with the Environmental Protection Agency (EPA) not having been established until 1970 (Pollack & Shaffer, 2009). The same year the United States passed the National Environmental Policy Act (NEPA) that enables federal agencies to take environmental concerns into account during decision-making processes. NEPA has often been used to challenge certain agency decisions regarding GM plant varieties.

According to Pollack and Shaffer, the Food and Drugs Act delegates the primary responsibility of food safety regulation to a federal body called the US Food and Drug Administration (USFDA). The USFDA is responsible for ensuring the safety of all food and food components, including the products of rDNA or GM technology, under the Federal Food, Drug and Cosmetic Act (FFDCA) (Nestmann, Copeland, & Hlywka, 2002). The USFDA has the authority to immediately remove any product from the market that poses a potential risk to public health or that is being sold without all necessary regulatory approvals. As a result, a legal burden is placed on developers and food manufacturers to ensure that foods available to consumers are safe and in compliance with all legal requirements of the FFDCA.

Nestmann, Copeland and Hlywka argue that in order to understand the regulatory approach of the USFDA regarding the safety evaluation of GM crops, it would be useful to analyse food and food safety from a historical perspective. It has been observed that people had been traditionally involved in producing and consuming agricultural food crops without any fear and concern over laws and regulations of food safety. Based on this evidence and experience, even today organic and conventional agricultural crops are accepted as safe for consumption without any additional safeguards to demonstrate their safety. However, the safety concerns have arisen only after GM crops came to the fore. This is because the GM crops are not being considered traditional crops, as they have no history and came into existence artificially through the work of scientists in the lab. Hence, people treat them as new crops requiring precautionary measures. Therefore, in order to analyse the precautionary measures needed to prevent any harmful effects related to food, federal regulatory bodies like FDA, NEPA and others have been formed.

Since the time the FDA was formed, it has been playing the leading role in approving the sale and marketing of GM foods in the United States. Parallel to the FDA regime, a second system that is labour intensive and rigorous in its functioning is in place, and that is the United States Department of Agriculture (USDA). The USDA was established in particular to regulate agriculture, including meat and poultry products. It was set up in 1862. It has not only helped in establishing food safety standards but also employs over 7,000 inspectors who carry out continuous inspection of meat and poultry plants. However, in the 1990s, the USDA came under severe criticism for its failure to detect and prevent outbreaks of *Escherichia coli* and other harmful bacteria in meat. This ultimately resulted in the adoption of some new safety procedures within the USDA in 1996.

Therefore, both the USDA and USFDA together act as the leading agency for granting safety approval for the release of food products into the market. Though these regulatory bodies – the USDA, USFDA and EPA – have been functioning for some time, they were not the first to evaluate safety guidelines of rDNA technology.

The National Institutes of Health (NIH) was the first federal regulatory agency to publish their interests in evaluating the safety of rDNA technology in 1976, in the form of guidelines for the conduct of research. Because of the uncertainties that existed at the time, the potential applications of rDNA technology and research studies were supposed to be conducted within the confines of federally funded laboratories under NIH control (Nestmann, Copeland, & Hlywka, 2002). After a series of continued research studies and careful assessments and monitoring of risks related to GMOs, the NIH released a set of less restrictive guidelines that were published in 1978. It further established an rDNA Advisory Committee (RAC) in the early 1980s to review all data and experience gained with the application of the technology under its control. Later in 1983, the NIH published a more relaxed set of guidelines based on the recommendations of the RAC. In that same year it approved the release of the first GMO developed through rDNA technology.

The NIH was severely criticised for this approval for failing to prepare assessment notes regarding the environmental impact of its regulatory decision as required under the NEPA. Later the NIH had to give up responsibility for the environmental assessment of GMOs. This came about mainly because Jeremy Rifkin, an anti-GM activist, sought and obtained an official order from the federal district court in Washington, D.C., against the NIH's approval of any release of GM varieties into the environment without the approval granted by the court (Pollack & Shaffer, 2009).

With all these visible facts and evidence about the working procedures of the US federal regulatory bodies, it was still unclear in which direction the United States would go in the first half of the 1980s: whether it would take a highly precautionary and process-based approach for GMO regulation or whether it would establish a uniform product-based regulatory approach was still a mystery then. However, the Reagan White House quickly responded to this mystery by forestalling a new legislation in Congress and forming a Biotechnology Science Coordinating Committee to resolve questions over jurisdiction, settle conflicting issues and help establish a regulatory product-based approach (Pollack & Shaffer, 2009). This further resulted in the curtailment of the EPA's role and the elevation of USDA and FDA roles. In 1986, the Office of Science and Technology Policy (OSTP) in the Reagan administration issued a 'Coordinated Framework' for the regulation of biotechnology, which continues to shape US biotech regulation even today (Pollack & Shaffer, 2009). The Coordinated Framework to regulate biotechnology thus formed a division of responsibility among the USDA, USFDA and EPA.

The European regulatory approach

The European regulatory approach towards GMOs can be described as a "complex system of interacting problems" (Tait, 2001). The European public has increasingly become negative towards GM crops, which is in deep contrast to the situation in the United States. There can be several reasons behind these differences in their approaches. In North America, agriculture is generally perceived to be an industry, whereas in Europe, it is more than an industry. According to Isaac (2002), agriculture in Europe plays a multifunctional role, such as supporting the rural way of life, preserving the culture and heritage of the countryside, ensuring the welfare of animals and protecting the environment.

According to Toke (2004), food is a major signifier of identity in Europe. Food in European society is crucially linked to a sense of belonging to a national community. The Americans, on the other hand, see themselves as the "melting point of many cultures". Unlike Europeans, cultural preferences in America have developed in such a way that most Americans prefer to eat out rather than at home. The working people there do not want to spend too much time preparing or cooking food. With the use of newly advanced technologies, they want every work, including cooking and manufacturing, to be ready within no time without putting in much effort. According to Toke (2004), EU countries can be associated with the newly emerging groups of the 'Slow Food Movement'. Toke argues that slow food is different from fast food. It can be defined as "to draw out a construction of what is good about quality food as opposition to bad production of corporate fast food". Therefore, as far as food is concerned, Europeans traditionally preferred quality to quantity. The quality that traditional food holds cannot be derived easily from today's GM technology. For example, "Roquefort cheese is an aromatic cheese and it has acquired this flavor gradually through traditional peasant life" (Toke, 2004, pp. 141–189).

GM foods have generally been perceived among Europeans as products born out of the latest technology but have no history and no specific place of origin. Toke argues that 'slow' or 'quality' food has links with the past and also has specific origins. According to him, slow food might not be necessarily organic food, but all organic foods easily slide into the realm of slow food. This is mainly because of the "organic appeal to tradition and also the ecological concern for the consumption of locally produced food".

In Europe, those foods which were never used for human consumption earlier are kept under the category of novel foods and novel food ingredients. The sale of all novel foods, including GM foods, is controlled by the EC Novel Foods Regulation. According to Tomlinson (2002), the main piece of European Community legislation that applies to novel foods is the EC Novel Foods and Novel Food Ingredients Regulation (258/97). It came into effect on 15 May 1997. Tomlinson argues that because of the introduction of these regulations, a harmonised pre-market approval process for a wide range of novel foods, including GM foods, was set. The sole purpose of adopting these regulations is to evaluate and describe procedures for assessing the safety of novel foods and specify labelling rules. From 1997 onwards, every year the structures of EC regulation were amended. EC Regulations no. 1139/98, no. 49/2000 and no. 50/2000 set out detailed rules for the labelling of food containing ingredients derived from GM soya bean and maize (Tomlinson, 2002). The EU's regulatory approach towards novel foods and ingredients could be understood more adequately if we analyse the European legislative framework that governs the release of GMOs into the environment.

The EU legislative framework for GMOs

The EU created a broad legislative framework in the form of directives for the deliberate release and marketing of GMOs in the 1990s (Tomlinson, 2002; Navneet, 2019). Among the many EU directives, Directive 90/220/EEC is of special importance because it is the first and exercises control over the deliberate release of GMOs into the environment. According to Tomlinson, the EU regulatory system is mainly set out in the form of EU directives to protect the environment and human health. It further aims to establish a single market in the EU for products containing GMOs. Therefore, Directive 90/220/EEC covers the release of GMOs for both experimental purposes as well as commercial sale.

According to the directive rules, all GMOs released to the environment, either for experimental or commercial purposes, must be assessed or approved by the Competent Authority (CA). For this purpose, the applicant has to file an application for the release of a GMO agricultural product, which has to be forwarded to that EU member state where the GMO-derived novel agricultural product is supposed to be placed in the market for the first time. The member state then refers the application generally to the CA for the assessment of the novel food product to detect whether it is safe for humans, animals and the environment. For adequate assessment, the CA can ask the applicant to provide all necessary details and information about the novel food as per the guidelines mentioned in the directive. In the case of insufficient information, the CA of the particular member state can further ask for more information from the applicant about the product as required. The applicant then has to provide compulsorily information statements assessing risks that the GMO would pose to human health and the environment. After the assessment work is done by the CA, other EU member states are invited to comment on the CA's decision. If the CA confirms that the information provided by the applicant ensures that the GMO would not pose any risk to human health and the environment, and if other member states have no objection to the assessment of the CA, then the GMO food product is approved by the EU commission to be placed into the EU market without any hindrance.

In case any member state raises an objection, the European Commission would submit the proposal to its Regulatory Committee, which is composed of government representatives of the EU member states. If the Regulatory Committee fails to deliver any suggestions, the commission further forwards the proposal to the EU Council of Ministers. If the council also fails to deliver any opinion within a period of three months, the commission is authorised to adopt its own proposed decision. Before doing so, the commission may go to the extent of asking its scientific committees for their opinion (Schauzu, 2011; Navneet, 2019).

Conclusion

Both the EU and the United States have been facing the common challenge of developing a regulatory framework for agricultural biotechnology starting from the early 1980s. After analysing their regulatory approaches, it can be argued that the two sides have adopted starkly different regulatory frameworks. The United States has adopted a product-based or science-based approach towards the use of biotechnology in manufacturing food. On the other hand, the EU's regulatory approach has been more precautionary, politicised and process based. However, according to Sprink, Erikson, Schiemann, and Hartung (2016), the discussions regarding how to accommodate genetically engineered techniques within the existing EU regulatory framework for GMO have intensified recently. As a result, the current regulatory system in the EU can be said to be both process and product oriented. But the regulatory mechanism of the EU has been, by and large, interpreted as strictly process based.

Pollack and Shaffer have argued that the existence of two distinctive and self-reinforcing regulatory frameworks in the United States and the EU can be deemed to be more than just of academic importance. This is because the domestic regulation of GM foods and crops in each side has external implications too. The impact of the EU's moratorium on GM food crops and its stricter regulatory standards was greatly felt in the United States, where around two-thirds of the GM crops are grown. This was the crucial cause of the trade dispute between the two giants of the world. The implication of this trade dispute was so great that it was felt in other parts of the world as well. For instance, today some poor African countries are in a confused position, as they are still pondering over which side to favour. An interesting fact to note is that in the US regulatory approach, food from GM crops is not considered to be fundamentally different from non-GM or conventional crops (Toke, 2004, pp. 96–140). This is because, according to Toke, US citizens have, by and large, a different attitude towards food in comparison to the Europeans. Americans seem to have greater trust in their regulators than do Europeans. With its product-based regulatory approach, the US regulatory system does not want to create any kind of distinction between GM crops and non-GM or conventional crops when both of them appear to be similar in phenotype. Therefore, there is no mandatory labelling done for GM food products in the market. The information on whether any food product is produced through GM technology would be provided to the American consumers only if they demand it separately.

However, in the EU, it is understood that most Europeans look at food in terms of quality cuisine. They want to know the quality of the food they are consuming and focus on the manner in which it has been prepared. Therefore, the EU countries have adopted process-based regulatory approaches. This means that there will be mandatory labelling for GM food products put into the market so that European consumers can decide whether they want to purchase GM or non-GM food products. The major concern in Europe about GM crops relates to health and environment issues.

In India, apart from health and environment as problematic issues, questions over ownership rights have also been raised. Here Bt cotton is the only GM crop that has been approved by the regulatory bodies for commercial cultivation. As a result, Indian farmers have increasingly become dependent on private seed companies like Mahyco-Monsanto to get new Bt cotton seeds, which sprout into plants whose seeds are sterile and cannot germinate and so cannot be preserved for sowing in the next season. This is a situation that the farmers are not used to in the conventional cropping pattern. The United States and the EU have a very explicit history of product-based and process-based regulatory stands. But in India it is still unclear whether the regulatory approaches are product based or process based.

Notes

1 Vandana Shiva. (1992). "The Seed and the Earth: Biotechnology and the Colonization of Regeneration", *Development Dialogue*. Retrieved November 14, 2013, from www.dhf.uu.se/pdffiler/9212/921-211.pdf
2 Of the two genes mentioned here, one would enable the seed to produce its own pesticide in order resist pests, and the second one would affect the embryo of the seed so that after growing into a plant, the seed does not germinate further.
3 Here Sahai is speculating that though today Monsanto might not be using terminator technology on Bt cotton in India, in the future it might use it to create a monopoly in the seed market in India. Her assumption is based on the fact that Monsanto has already bought two companies, Delta and Pineland, who were co-owners along with the US government of the terminator technology.
4 Suman Sahai. *Genetically Modified Crops: Issues for India*. Retrieved January 4, 2012, from www.genecampaign.org/Publication/Article/GMtech/Genetically-ModifiedCrops.pdf
5 Vandana Shiva. (1992). "The Seed and the Earth: Biotechnology and the Colonization of Regeneration", *Development Dialogue*. Retrieved November 15, 2013, from www.dhf.uu.se/pdffiler/9212/921-211.pdf
6 Ibid.
7 See talk by Suman Sahai at "India Today" conclave 2010. Retrieved January 23, 2012, from www.youtube.com/watch?v+vy4W4v-HPiY&feature+youtube.be
8 The Genetic Engineering Approval Committee (GEAC) is one of six competent regulatory authorities in India that authorises large-scale trials and environmental release of GMOs. It falls under the Ministry of Environment and Forests (MOEF).
9 Gargi Parsai. (2013, April 27). "Global Scientists Back 10-year Moratorium on Field Trials of Bt Food Crops", *The Hindu*, S&T Agriculture.
10 MoEFCC is the new name of MoEF.
11 See the Hindu newspaper dated 3 August 2013 in the national news section. Nitin Sethi is reporting on "Jayanthi Natarajan Opposes Pawar's Views on GM Crops, Wants Field Trials Put on Hold". Retrieved December 25, 2016, from www.thehindu.com/news/national/jayanthi-natarajan-opposes-pawars-views-on-gm-crops-wants-field-trials-put-on-hold/article4982776.ece
12 The BRAI bill has lapsed in the parliament.

13 Ibid.
14 See the *Hindustan Times* newspaper dated 28 February 2014 in the India news section. Chetan Chauhan reports the news titled "Veerappa Moily Clears Field Trials of GM Crops". Retrieved December 25, 2016, from www.hindusta-ntimes.com/india/veerappa-moily-clears-field-trials-of-gm-crops/story-Sgps5nLz9P6AtoFven4Y6H.html
15 See the *Business Standard* newspaper of 26 and 30 March 2014.
16 The Alliance for Better Foods (ABF) is run by the Washington, D.C., office of BSMG Worldwide, a full-service PR firm whose clients include Monsanto, the Chemical Manufacturers Association, Procter & Gamble, Phillip Morris and numerous other large food, chemical and pharmaceutical companies. The ABF was created mainly to promote public acceptance of GM foods and oppose their labelling in order to distinguish them from normal or non-GM foods. For more information, please visit www.sourcewatch.org/index.php/Alliance_for_Better_Foods and www.betterfoods.org/ websites.

2 Critical analysis of India's regulatory approach

Introduction

It is a well-acknowledged fact that agriculture and industry are two important economic sectors in India. Today both these sectors are being exposed to a new set of technological developments called biotechnology. India is one of the first Asian countries to invest in agricultural biotechnology research. From the Indian point of view, the agricultural sector is very important for bringing inclusive growth to the Indian economic system. To increase agricultural growth, the new technology that is being considered for use is genetically modified technology.

According to George T. Tzotzos, Graham P. Head, and Roger Hull (2009), "Genetic Manipulation is a technology in which a gene or genes are taken from one organism (the donor) or are synthesized or modified and then inserted into another organism (the recipient) in an attempt to transfer a desired trait or character" (Tzotzos, Head, & Hull, 2009, p. 13). However, this technology is not free from controversies and complications that might affect the lives of human beings and other living creatures on earth. To deal with these issues, the Indian government has set up a biosafety commission to regulate the approval of genetically modified crops/foods. In the introductory chapter of the book, it was shown how the United States and the EU adopted product-based and process-based regulatory approaches, respectively, for GM technology. In that context, it would be interesting to look at the historical background of the regulatory approach adopted by India for the same technology.

Therefore, the basic aim of this chapter is to analyse various policies and politics that came into shape with the arrival of GM crops in India and the manner in which the processes of the Indian regulatory system developed for evaluating the safety concerns over genetically manipulated or modified crops. Apart from this, the different debates and critical arguments of scholars relating to the Indian regulatory approach will be considered in order to understand the controversy surrounding the Indian regulatory system.

The legal regulatory framework in India

The basic legal framework governing novel foods (both GM and non-GM food products) in India is the Environmental Protection Act (EPA) of 1986. In 1989, the Government of India (GOI) formulated the rules and guidelines for the manufacture, use, import, export and storage of hazardous micro-organisms and genetically engineered organisms (Kolady & Herring, 2014). These rules became effective from 13 September 1993 and constituted the legally binding regulatory framework for GMOs and other novel foods in India (Anuradha, 2005).[1] According to Gupta (2000), in accordance with the 1989 guidelines, GMOs are placed in the category of hazardous micro-organisms, and their regulation under the EPA is justified by their alleged potential to be environmental pollutants. According to Anuradha R.V. (2005), sections 6, 8 and 25 of the EPA are of special importance, as they give power to the central government to deal effectively with environmental pollutants. Section 6 of the act refers to the rules to regulate environmental pollution and specifically mentions the need for procedures and safeguards to handle the hazardous substances.[2] Similarly, section 8 mandates that any person handling hazardous substances is bound to comply with the pro-cedural safeguards. In addition, section 25 grants the central government power to make rules to maintain safeguards and adopt a precautionary approach against any new additives in human or animal food products. These additives may be either GM or non-GM.

The EPA guidelines of 1989 do not mention separately anything about GM additives in food products. They seem to group GMOs with hazard-ous substances. Therefore, it became necessary for the government and the regulatory body to revise these guidelines later in the 1990s; by this time, GM technology had become important in both the agricultural and phar-maceutical sectors. With the enhancement of GM technology in the 1990s, it became clear that India might become an important market for the sale and purchase of GM products in the future. Therefore, the 1989 guidelines of the EPA were revised twice: first in 1990 and again in 1998.

In simple terms, in India, GMOs are regulated under the EPA of 1986 as well as the Indian Biosafety Regulatory Framework. This framework con-sists of several rules and guidelines. They are the 1989 "Rules for the Manu-facture, Use, Import, Export and Storage of Hazardous Micro-organisms, Genetically Modified Organisms and Cells"; 1990 DBT's "Recombinant DNA Safety Guidelines"; 1994 DBT's "Revised Guidelines for Safety in Bio-technology"; and 1998 DBT's "Revised Guidelines for Research in Trans-genic Plants and Guidelines for Toxicity and Allergenicity Evaluation of Transgenic Seeds, Plants and Plant Parts".[3]

Consent givers in the biosafety regulatory framework

India's GMO governance can be compared to that of the European regula-tory framework, as both India and the EU have adopted a precautionary approach towards the use of GM food products, which is in line with Codex Alimentarius.[4] Nevertheless, the Indian government's approach can be con-sidered a lot different from that of European and American governments. This is because the Indian government maintains very close control of the different applications that are under process. In the United States and the EU, it is not the government, but the regulatory agencies that are involved with the applicant or the sponsor (Salat, Salter, & Smets, September 2010). It has been observed that most of the research studies in India are funded through government initiatives, notably from the Department of Biotechnology (DBT). Further, in the Indian regulatory system, it has been observed that before GM food products can be marketed, the applications for their approval have to pass through six competent authorities (Shukla, Al-Busaidi, Trivedi, & Tiwari, 2018). These six competent authorities are:

1   The Institutional Biosafety Committee (IBSC) – This monitors the projects at the R&D level.
2   The Review Committee on Genetic Manipulation (RCGM) – This monitors research activities and small-scale field trials.
3   The Genetic Engineering Approval/Appraisal Committee (GEAC) – This authorises large-scale trials and the environmental release of GMOs.
4   The Recombinant DNA Advisory Committee (RDAC) – Under DBT, it prepares adequate and suitable safety recommendations and reviews developments achieved in this regard at the national and international level.
5   The State Biotechnology Coordination Committee (SBCC) – This coordinates and monitors GMO activities in the state with the central ministry.
6   District-level committee (DLC) – This monitors GMO activities at the district level.

It has been observed that out of these six competent authorities, only the first three or four are actively involved in the process of granting approval to the GMOs.

The "EPA Rules 1989" define the six competent authorities as responsible for the regulation of GM crops under a three-tier system. The first tier is the Advisory Committee and includes RDAC. It comes under the pre-research domain as an advisory body to facilitate and promote research. The second tier is the Approval Committee and includes the IBSC, RCGM and GEAC. They function under the research domain, closely monitoring research and experimental releases, such as field trials. The third tier is the Monitoring-cum-Evaluation Committee (MEC) and includes SBCC and DLC. They are mostly the part of the post-release domain (Kolady & Herring, 2014).

Gupta (2000) argues that in pursuance of the 1989 Biosafety Rules, the authority to regulate GMOs is divided between the Ministry of Science and Technology and the Ministry of Environment and Forests (Gupta, 2000, p. 15). The DBT comes under the Ministry of Science and Technology. One of the most important organs of the DBT is the Review Committee on Genetic Manipulation (RCGM). Research studies on GMOs are mainly overseen by the RCGM. It consists of members of the DBT, Indian Council of Medical Research (ICMR), Indian Council of Agricultural Research (ICAR) and the

Council of Scientific and Industrial Research (CSIR). The committee is also responsible for creating guidelines specifying the procedure for the regulatory process and for ensuring adequate precautions during field trials (Salat, Salter, & Smets, September 2010). According to Anuradha R.V., RCGM can also give approval for controlled field experiments, but these would be conducted only on a small scale (Anuradha, 2005, p. 29). To maintain transparency in the working system of the RCGM, its ongoing project and field trial data are published regularly on the Indian GMO Research Information System (IGMORIS) website.

The *Institutional Biosafety Committee (IBSC)* is involved in reviewing and giving clearance to those project proposals that fall under any of the categories mentioned in the 1989 rules. It is mandatory for all research institutions to set up this kind of committee, which would be composed of scientists, a medical expert and a nominee of the DBT (Salat, Salter, & Smets, September 2010, p. 51). Like RCGM, IBSC also comes under the DBT. Further, RCGM is given updates on the various project reports that are underway in each IBSC of various research institutions.

The *Genetic Engineering Approval/Appraisal Committee* (GEAC) falls under the Ministry of Environment and Forests (MoEF). It is responsible for the approval of activities involving large-scale use of GMOs and their release into the environment through experimental field trials (Anuradha, 2005). In other words, the deliberate release and commercialisation of GMOs is to be overseen by the GEAC. Like RCGM and IBSC, GEAC is composed of representatives from the MoEF, DBT, ICAR, CSIR and ICMR, among other experts.

The *Ministry of Environment and Forests* (MoEF) is also in charge of applying the Cartagena Protocol in order to protect biological diversity. The Cartagena Protocol on Biosafety is an international agreement on biosafety that is a supplement to the Convention on Biological Diversity. The Biosafety Protocol makes clear that products from new technologies must be based on the precautionary principle and that they should allow developing nations to balance public health against economic benefits.[5] India is one of the few countries that exports GMOs in the form of Bt cotton and still has ratified the Biosafety Protocol. So far, India has not faced much complication in its GMO exports, which are governed by the protocol it has signed, because it is still not a major GMO exporter. However, the situation might change in the future when more Indian GM food products reach the international market. According to the GEAC policymakers, India ratified the Biosafety Protocol with the sole purpose of doing prior research and controlling the intentional release of GMOs that might harm public health and disturb the environmental balance (Salat, Salter, & Smets, September 2010, p. 51).

According to Gupta (2000), the functions of these two national Biosafety Committees, i.e. RCGM and GEAC, appear to be clearly delineated. But the division of responsibility among them still remains a source of much controversy. Disputes have centred on the boundary between research and deliberate release of GMOs, and in particular, whether field trials constitute a research activity or a deliberate attempt to release GMOs into the environment (Gupta, 2000, p. 15). According to Gupta (2000), there is great bewilderment over the issue. This is because if the field trials are deemed to be 'research', then according to the guidelines or rules of EPA, they should be regulated by DBT. But if they are deemed to be the 'deliberate release of GMOs into the environment', then they should fall under the regulatory domain of GEAC of MoEF.

The *State Biotechnology Co-ordination Committee (SBCC)* is to be constituted at the state level. It is responsible for performing periodic reviews of safety and control measures adopted by various industries and institutions handling GMOs. The SBCC functions under the supervision of the GEAC at the MoEF (Anuradha, 2005, p. 29). Similarly, the *DLC* is to be constituted under the district collectors in every district to monitor the safety regulations of industries dealing with GMOs.

The 1998 guidelines introduced a seventh committee called the *Monitoring and Evaluation Committee (MEC)*. It is authorised to conduct field visits at experimental sites, collect data and advise the RCGM on risks and benefits. It can also recommend the required changes and remedial measures in order to bring improvements in the field experiments (Anuradha, 2005, p. 29).

Critical analysis of Indian regulatory bodies

Carl E. Pray, Prajakta Bengali, and Bharat Ramaswami (2005), in their article "The Cost of Biosafety Regulations: The Indian Experience", have analysed the overall financial costs involved in the regulatory rules followed by the public and private sectors. They argued that for approving GM food crops in India, the regulatory system requires more and more data to meet the regulatory requirements. In addition, a new GM crop will have to go through two years of multilocation field trials in the all-India coordinated trials maintained by the ICAR. These multilocation field trials would cost $1,000 per site (Pray, Bengali, & Ramaswami, 2005). Up to now in India, no GM food crop has been approved. According to the article, private companies speculate that food crops such as maize or soya bean will require much more testing and time for approval compared to non-food GM crops such as Bt cotton. The three authors in their article have shown that the regulatory mechanism of GM technologies provides both incentives and disincentives to do research. The incentive effect of regulation comes from the intellectual property rights (IPRs) on specific transgenic events that companies or research institutes are able to obtain. The disincentive effect comes from the regulations that increase the cost for bringing a new GM product into the market. The incentive and disincentive effects can be understood more broadly and adequately by analysing the question why GM rice and other GM food crops have failed to get approval for commercial cultivation, whereas Bt cotton easily managed to get such approval. According to Pray, Bengali, and Ramaswami (2005), the transgenic rice has been approved nowhere in the world for commercial cultivation. Therefore, in the absence of a proper dossier of information about transgenic or GM rice, the regulatory costs would naturally go higher. Besides, India is known to be the centre of biodiversity for rice, and therefore has a special ecological concern for the crop. Such concerns are not relevant in the case of cotton. For all these reasons, private firms in India estimated that it would cost around $1 million or even more to obtain regulatory approval for GM rice than for GM cotton (Pray, Bengali, & Ramaswami, 2005). In addition, rice is deemed to be the most important food grain in India. It is natural that the Indian regulators should adopt more precautionary measures in order to check food safety risks related to GM rice. All these factors lead to considerable uncertainty about when and whether GM rice would ever receive approval in India.

There is speculation by some scholars that with the approval of GM seeds by the Indian regulators, new problems are bound to arise, such as illegal transportation of some GM seeds to the neighbouring countries, where an adequate regulatory system to examine the risks associated with GM seeds still does not exist. GM seeds have still not been approved by the government bodies in these neighbouring states. In order to stop such practices, the Indian state and its regulatory bodies would also have to take its neighbouring states into confidence through proper channels. Some private organisations have already started working in this regard. For instance, a private company called "Bayer has started seeking approval for the transformation events in other countries even though the target market is India" (Pray, Bengali, & Ramaswami, 2005).

Lianchawii, in her article "Biosafety in India: Rethinking GMO Regulation", has strongly criticised the inadequacy and inefficiency involved in the Indian regulatory structure, especially regarding an incident in Gujarat. According to her, poor coordination, breakdown of communication between the centre and the state authorities and absence of monitoring agencies were some of the glaring weaknesses that were observed in the system. For instance, the GEAC approved Bt cotton cultivation in Andhra Pradesh despite the absence of the SBCC and the DLC (Lianchawii, 2005). As already discussed, both SBCC and DLC were proposed to oversee the implementation of the regulation as well as the performance of GM crops. Lianchawii reports in her article that when several field trials of Mahyco's Bt cotton were on, SBCCs had not yet been formed in most of the Indian states.

Suman Sahai, in her article "Can India Handle GM Technology?", writes about the hiccups when in November 2002, GEAC convened a meeting to grant approval to ProAgro's GM mustard varieties. Some NGOs like Gene Campaign made several allegations against the Indian regulatory body that the GM mustard had not cleared all the required safety tests and that it had not been sufficiently tested in the fields. Despite Gene Campaign's demand for the release of the field data presented by ProAgro, a private company, the

GEAC was reluctant to release the complete field data. Further, the regulatory body also did not go for separate field trials to understand the relevance and adequacy of the data presented to it by the private company. Because of the lack of transparency in Indian regulatory bodies, NGOs and farmers have been demanding that the company be required to do new field tests that can be monitored by some independent experts.[6] The director general of ICAR in 2001,[7] Dr Panjab Singh, said that ProAgro's mustard data are not sufficient to justify the commercial release of GM mustard. He suggested that additional tests would have to be done, especially because most of the data provided to the GEAC were provided by ProAgro itself.[8] This shows the clear conflict of interest. The regulatory bodies are intending to approve GM crops blindly merely on the data provided by the private companies, which raises questions on the transparency and accountability of the regulatory bodies. According to some scholars, a major loophole has shown up in the 1989 rule itself, which was framed to regulate adequately GM technology for the benefit of human beings. For instance, the 1989 rules say that in cases where GM technology uses do not comply with the rules of regulatory bodies like DLC or SBCC, measures should be taken against that person or organisation responsible for not complying with the rules. But the rules do not provide for any penalty such as confiscation or compensation that could deter such prospective offenders.[9] In addition, though no DLC or SBCC has been made functional as of now, in some Indian states GM crops like Bt cotton are already being commercially cultivated.[10] This proves that there is a large lacuna in making the laws and enforcing them.

Ashok B. Sharma writes that for speedy approval of GM crops, then Indian Science and Technology Minister Prithviraj Chavan intended to push through a bill in parliament to set up a new National Biotechnology Regulatory Authority. Chavan's proactive stance came only after the denial of Bt brinjal (eggplant) by his colleague Jairam Ramesh, the then environment and forest minister, in the face of widespread public protest against its approval by the GEAC. "The environment minister is the person who has the discretionary power to withhold the approval of GEAC, which after all is a body of handpicked scientists many of whom have conflict of interests".[11] After the introduction of GM crops whose danger to health and environment became widely recognised, world leaders came together to express concern and ratified the Cartagena Protocol on Biosafety, which came into effect on 11 September 2003 (Sharma, 2010). India has not only signed and ratified this global treaty but is also an active participant in it. Sharma (2010), in his article, recommends that Science and Technology Minister Chavan should have placed more stress on setting up a National Biosafety Authority rather than a National Biotechnology Regulatory Authority and that he should have become thoroughly aware of the country's position on the Cartagena Protocol. Article 27 of the protocol recommends that all countries set up a global liability and redressal mechanism for damages caused on account of transboundary movement of GMOs, also living modified organisms (LMOs).[12] In India so far, Bt cotton is the only GM crop that has been approved. The GEAC has been cautious in not approving any GM food crop. According to Sharma (2010), the GEAC under the chairmanship of A.M. Gokhale denied approval to three GM hybrid mustard seeds developed by ProAgro in collaboration with Aventis and PGS of Belgium for commercial cultivation. Sharma further states that during Gokhale's tenure, the GEAC also refused imports of hazardous Star Link corn. But Gokhale could not continue as GEAC chairperson for long, as pro-GM lobbies ensured his exit (Sharma, 2010). This shows the extent of the influence of private players in government-controlled regulatory bodies and further illustrates the nexus between government and private companies to promote GM crops without much concern for the safe handling of GMOs and public health.

Such instances of nexus can be clearly understood if we study the analytical writings of Peter Newell on corporate strategies to promote GM crops in the markets of a developing country like India. Newell (2003) writes that the period from 1987 to 1989 laid the foundations for the development of a strong private-sector seed industry in India. The economic liberalisation launched in 1991 served to hasten this process. By loosening restrictions on the activities of foreign firms and multinational companies and by giving automatic approval to foreign technology agreements, the Indian government only facilitated private-sector plant breeding. As a result, the private-sector investment in the seed sector in India more than tripled between 1993 and 1997 and reached a high of Rs 19,850 million (Newell, 2003). Manju Sharma, the then DBT secretary, claimed that the new seed policy would directly promote the seed industry and suggested that the sector is now poised for a quantum jump due to the changes made in the regulations (Newell, 2003). Similarly, Mahyco's joint director of research and development, Dr Usha Zehr, claims that India has the potential to position itself as a key market for biotechnology products, as the European and North American markets have reached a saturation level (Newell, 2003).

Some multinational companies in India with large biotech portfolios are Monsanto, Syngenta, Du Pont and Aventis. Among these, Monsanto is the most high-profile and maintains the largest presence in India. It has a research centre in Bangalore at the Indian Institute of Science (IISC), which has often become a magnet for various protests against GM development (Newell, 2003, p. 4). Newell, in his article, has explained at great length how astutely Monsanto managed to get approval for the commercial release of its Bt cotton. According to Newell (2003), in 1990 when Monsanto sought approval for Bt cotton release, its proposal was rejected by the regulatory bodies on the basis of the high cost of technology transfer. However in 1995, Mahyco, the long-established Indian seed company headed by Dr Barwale, was granted permission to import 100 grams of transgenic cotton seed as part of an agreement with Monsanto (Newell, 2003). Taking full advantage of this opportunity, Monsanto further consolidated its position by buying a 26 per cent stake in Mahyco in 1998, thus creating Mahyco-Monsanto Biotech India, Ltd. (MMB) (Newell, 2003). This was an astute and strategic move by Monsanto, as Mahyco's director, Dr Barwale, is known to be very influential in the Indian regulatory system. He is a well-respected member of the Indian agricultural industry and has been honoured by the Indian government for his contributions to the agricultural sector. According to Newell, his connections within the government extend beyond the DBT to many of the key agencies involved in biosafety regulations (Newell, 2003). Soon after the linking up of Monsanto with Mahyco, in March 2002 MMB's Bt cotton was approved for commercial release for a three-year trial period in six states (Newell, 2003). Newell further states that Monsanto's position in India has become so strong that it now maintains its own 'regulatory affairs' office in Delhi and engages in routine interactions with government officials over policy development. It has led attempts to get media publicity by funding public survey demonstrations in support of GM technology and has funded some selected rich farmers with large landholdings to attract other farmers in favour of GM crop development (Newell, 2003).

According to A. Damodaran, both corporate industries and civil society wanted structural changes in the biosafety regulations in India. However, the solutions proposed by private and corporate industries, on the one hand, and NGOs and civil society groups, on the other, are radically different (Damodaran, 2005). While the first one, i.e. the private or corporate industries, aim to "short-circuit the regulatory process", the latter, i.e. NGOs or civil society groups, seek elongation of the regulatory processes to the maximum extent through inclusive and participatory approaches (Damodaran, 2005).

The two perspectives, i.e. the corporate sector's and the civil society's, can be understood more broadly if we analyse the papers of different scholars who have taken either of these positions. For instance, Jennifer Ifft in her paper has illustrated the point of view of the biotechnology industry, while scholars like Sahai and Kuruganti have written from the perspective of civil society and farmers' rights.

In her paper, Jennifer Ifft (2001) talks about the regulatory confusion related to GM crop approval in India. According to her, the GOI has often been criticised for channelling the arbitrary and confusing process of GMO approval (Ifft, 2001, p. 5). However, according to her, it should not be a big surprise, as the inefficiency of bureaucracy is no great secret. The All India Biotechnology Association (AIBA) released a report in November 2000 blaming the complicated bureaucratic process for the failure of India to adopt GMOs adequately (Ifft, 2001). The report covered all the inadequacies of GMO regulation carried out by the Indian regulatory bodies. According to the AIBA report, out of six regulatory bodies, only two, i.e. RCGM and GEAC, have been actively involved in the majority of decisions related to GM crops (Ifft, 2001). The report further shows that although the make-up of these two regulatory bodies, i.e. RCGM and GEAC, is similar in some respects, their decisions vary most of the time. For instance, RCGM has a

DBT representative, and similarly GEAC too has one, who co-chairs GEAC committee. RCGM has an ICAR representative, and one of the expert members of the GEAC is the director general of the ICAR. Despite the presence of some representatives from the same organisations in both regulatory bodies, GEAC frequently delays approval for the commercial release of GM crops that have already been approved by RCGM (Ifft, 2001).

Therefore, the distinction between small-scale and large-scale trials has drawn criticism from corporate and private biotechnology industries. The AIBA, along with the Confederation of Indian Industry (CII), recommends bringing an end to such distinction by single-window clearance for transgenic crops. According to them, the simplified regulation would be an improvement and speed up the implementation of biotechnology in India.

Taking into consideration this core interest of the corporate sector involved in biotechnology research, the DBT drafted a new revised bill called "Biotechnology Regulatory Authority of India" (BRAI) in 2009 (Salat, Salter, & Smets, September 2010). BRAI was supposed to provide a single-window mechanism for biosafety clearance, and the Ministry of Health and Family Welfare (MHFW) would have provided the administrative support (Salat, Salter, & Smets, September 2010). However, the bill was never discussed and lapsed in parliament. It was expected that until the bill on BRAI became law, the existing regulatory mechanism under EPA 1986 and Rules 1989 would continue to be in force. It was proposed that BRAI should initially have three regulatory branches (Salat, Salter, & Smets, September 2010):

1   Agriculture, Forest and Fisheries Branch (AFFB)
2   Human and Animal Health Branch (HAHB)
3   Industrial and Environmental Applications Branch (IEAB)

The Centre for Indian Political Research and Analysis (CIPRA) has critically analysed the draft of BRAI formed by DBT and reports that according to the new bill, BRAI would not come under the purview of the Right to Information Act 2005. It would also have authority to override the decisions of state governments, thus reducing the state authority to that of a mere spectator when it comes to making decisions on introducing GM food.[13] According to the CIPRA report, BRAI would be empowered to bypass the provisions enumerated in the Biological Diversity Act as well as the EPA. CIPRA criticises the bill by arguing that it apparently aims to lower the bar for introducing GM crops without taking into consideration their impact on the health of people and the environment. If the BRAI bill is passed in parliament, it would deprive people of their right to grow, own, trade, transport, share, feed and eat food that nature has bestowed upon humanity. Farmers would not have control over what they grow, and the public would not have control over what they eat. All controls would rest with the profit-hungry agro-corporate sectors, and the common people would be reduced to an expendable pawn at the hands of capitalists and their agents.[14] According to Avik Roy, the way this new bill has been given shape simply shows the blatant nexus between multinational seed companies who are pushing their GM crops in India and our government institutions.[15] Roy further states that the bill has been widely criticised for its undemocratic and promotional approach towards GM crops instead of a precautionary one. He further argues that the bill seems to offer a monopoly in the Indian agricultural market for multinational seed companies and private seed industries. According to the report presented by the NGO Greenpeace, the bill simply makes a mockery of biosafety assessment. It does not have a clause or section at all for rolling back approvals, or labelling GM food or deterring in any way the liability of the crop developer due to economic losses caused by contamination or other means.[16] Also, it has been observed that the BRAI bill did not adopt any of the recommendations given by the task force under the chairmanship of Dr M.S. Swaminathan, which was formed under the Ministry of Agriculture in May 2003.

The task force was formed mainly to examine potentialities and problems involved in biotechnology applications. After studying the situation, it recommended to set up an independent and professional watchdog, the National Biotechnology Regulatory Authority (NBRA) (Damodaran, 2005). NBRA would mainly aim to generate public confidence in the use of GMOs. The task force suggested that until the NBRA is formed, the GEAC would look into the matter of biosafety and environmental safety. The MEC would report to GEAC all issues concerned with biosafety and environment safety.

Sahai (2004) has thrown some light in her commentary article in the journal *Current Science* on the recommendations made by the Agbiotech Task Force committee headed by M.S. Swaminathan. According to her, it is an important step forward in trying to improve the system for implementing agbiotechnology in India. She argues that its importance lies in the fact that it is the first recommendation of its kind for change from a high-powered source and the first effort to formulate a policy (Sahai, 2004). Earlier civil society organisations and NGOs had been frustrated because of DBT's recalcitrance and refusal to engage in any dialogue on public concerns and its disregard for any suggestions for improving a clearly unsatisfactory system. Sahai says that the former head of the DBT has famously been on record for doggedly insisting that India does not need any biotechnology policy at all at a time when all around, from top scientists to the most vocal protagonists and opponents, saw the seriousness of the matter and were demanding a national policy (Sahai, 2004). According to Sahai (2004), the task force report contains many positive features. Its most basic and essential recommendation is that "the national policy should seek the economic well-being of farmers' families, food security of the nation, health security of the consumer, protection of the environment and the security of our national and international trade" (Sahai, 2004). The task force report further recommends that alternatives for GM technology be adequately examined and used only in the absence of other options. It is critical in particular of the prevailing gung-ho climate when any proposal for the research of a GM crop, however nonsensical, gets sanctioned, often at the cost of solid and conventional research. It also recommends that transgenic research not be done on those Indian crops that have a good demand in the international market and add to the foreign exchange. These crops are soya bean, basmati rice and Darjeeling tea (Sahai, 2004). One of the major reasons to adopt such a recommendation is because India's major trading partners have all rejected GM foods (Sahai, 2004). According to Sahai (2004), the committee report further suggested that the government policy on transgenics be sensitive to biodiversity conservation and the socio-economic context of our composite agrarian system. It further recommends protecting farmers' interests and that they should have the right to save seeds from previous harvests.

The Indian Council of Medical Research in New Delhi also published a paper analysing the Indian regulatory approaches towards GM crops at great length and gave several recommendations for their improvement and smooth functioning. According to ICMR's analysis, a comprehensive safety assessment of GM foods needs to be carried out in harmony with the Codex Alimentarius Commission and Cartagena Protocol (ICMR, 2004). The Health Ministry needs to have a special committee on novel foods as part of the Central Committee for Food Standards (CCFS), to be chaired by the director general of ICMR. All applications, including labelling of GM crops received by the RCGM/GEAC, need to be referred and cleared by this committee.[17] The article published by ICMR further reveals that policies on labelling of GM foods differ from country to country, but the consumer's right to know to enable them to make an informed choice is generally recognised. According to the ICMR (2004) article, labelling is compulsory in the EU, Japan, Switzerland and Australia. However, in the United States there is no such obligation to label GM foods unless there are substantial changes in the food crop composition. By analysing the labelling mechanism in other countries, ICMR analysts present a case for India and argue that in India there is a need to compulsorily label a food product if it contains novel DNA/protein or has altered characteristics. However, the ICMR analysts also recommend that it is not necessary to label some foods like refined edible oil and food ingredients added in minor quantities. Further, the imported foods are also subjected to label regulations (ICMR, 2004). They should have a certificate of origin indicating their GMO status and proof of analysis from some certified laboratories.

Conclusion

It has been observed by scholars like Sahai, Kuruganti, Lianchawii, Damodaran and several others that the Indian regulatory system to date has not implemented any recommendation and suggestion given by committees like the Agbiotech Task Force and ICMR analysts. Kuruganti has blamed the Indian government for apparently acting in response to influence and pressure from the United States to open the Indian markets for GM crops developed in the United States. According to Kuruganti, on 23 May 2006, India received a notification from the United States through the WTO committee on "Technical Barriers to Trade", which expressed American concerns, reservations and objections on India's move to label and certify GM foods (Kuruganti, 2006). She has further argued in her article that after Bayer's GM mustard was turned down in 2002, India is again on the verge of approving Bt brinjal, another GM food crop, for large-scale trials. She reminds her readers that at the level of large-scale field trials, it is mostly agronomic evaluation that counts. This is because biosafety tests have already been completed at small-scale field trials or at the RCGM level. She further recalls that it was at the field trial stage that the first discovery of the contamination of Indian cotton with illegal Bt cotton was made in Gujarat in 2001. Since then, the unbridled proliferation of illegal Bt cotton in the country has been proof of serious regulatory failure. Therefore, there exists much concern that these large-scale field trials should not become synonymous with commercial cultivation permission too.

In India, brinjal is one of the most important food crops. Farmers grow them in different varieties here. Sahai and Kuruganti argue that India never faced any crisis in growing brinjal. Therefore, like GM rice and GM mustard, introducing Bt brinjal in India is also questionable. The major farmers' organisations in the country, including the All India Kisan Sabha, Bhartiya Kissan Union, Bharat Krishak Samaj, Shetkari Sangathan, Andhra Pradesh Rythu Sangam, etc., have questioned the need to introduce Bt brinjal and other GM food crops in the country (Kuruganti, 2006). But the manner in which the regulatory bodies have been operating in India has compelled several civil society groups to ask: "Why should there be such great haste in giving approval to GM crops?"

It has been observed that though on paper the application for GM technology crops has to pass through six competent regulatory authorities for the approval of field trials, in reality, only three of the six bodies are functional. They are IBSC, RCGM and GEAC. SBCC and DLC, which are supposed to be functional at the state and district level, respectively, to monitor GMO activity, are dysfunctional in most of the states. Therefore, there is an urgent need to make these regulatory bodies active at the state and district level, as well for the effective monitoring of GMOs in agriculture.

However, it has been observed that AIBA and CII have been lobbying the government to create a single-window system for speeding up biosafety clearance for GM technologies. The UPA government tried to implement this by drafting a bill to set up a BRAI in India. However, the bill has lapsed in the absence of consensus over it in the parliament. It has been observed that Monsanto, the private company that holds patent rights for GM technology, has a strong influence with the government. Initially, when Monsanto sought approval for the commercial release of its Bt cotton seed varieties from the Indian regulatory bodies, its proposal was rejected. Thereafter

Monsanto very strategically formed a partnership venture with the Indian seed company Mahyco to launch Bt seeds in the Indian market. Mahyco's director, Dr B.R. Barwale, was deemed to be an influential person in the Indian regulatory system. As a result, soon after Monsanto's linking up with Mahyco, Bt cotton was approved for commercial release in six Indian states.

It needs to be considered that India is a member of the Codex Alimentarious Commission (CAC) and has also signed the Cartagena Protocol on Biosafety (CPB) to the Convention on Biological Diversity (CBD). The CAC provides guidelines and standards for food safety. Similarly, CPB is an international agreement aimed at ensuring adequate protection for the safe transfer, handling and use of LMOs resulting from modern biotechnology experiments and uses, which may have an adverse effect on the conservation and sustainable use of biological diversity and human and animal health. Therefore, unlike the United States, India has adopted a precautionary approach for the regulation of GM technology.

Though everything seems to be fine on paper, lacunae remain in the implementation of the rules by the regulatory bodies. The ICMR has proposed the establishment of a special committee, such as a CCFS, to take care of the safety assessment of GM crops in harmony with the guidelines of CAC and CPB. ICMR has also recommended mandatory labelling of GM food crops before granting any approval to them. But so far, the regulatory bodies have not put in place guidelines for an adequate labelling mechanism for GM food products. This shows that there is lack of seriousness among the regulatory bodies for assessing GM technology approvals.

Notes

1 R.V. Anuradha. *Regulatory and Governance Issues Relating to Genetically Modified Crops and Food: An India Case Study*. Retrieved May 12, 2012, from www.google.co.in/search?source
    Aarti Gupta. 2000. *Governing Biosafety in India: The Relevance of the Cartagena Protocol*. Belfer Center for Science & International Affairs, Global Environmental Assessment Project: Environment and Natural Resources Program, Harvard University, p. 13. Retrieved May 14, 2012, from glogov.org/images/doc/Gupta2000.pdf
2 For details refer *The Environment (Protection) Act, 1986*. Retrieved from www.envfor.nic.in/legis/env/eprotect_act_1986.pdf
3 For details refer *GMO Regulations in India and Their Weaknesses*. Retrieved the "*Gene Campaign*" website from www.genecampaign.org/Publication/Article/gmo-reg-india-weakness-p1=ID1.htm
4 The Codex Alimentarius is recognised by the World Trade Organization as an international reference point for the resolution of disputes concerning food safety and consumer protection.
5 Refer to *Cartagena Protocol on Biosafety*. Retrieved May 29, 2012, from http://en.wikipedia.org/wiki/Cartagena_Protocol_on_Biosafety
6 Refer to the article of Suman Sahai. *Can India Handle GM Technology?* Retrieved June 10, 2012, from www.genecampaign.org/Publication/Article/GMtech/CanIndiaHandleGMTechnology.pdf

7   See the Businessline newspaper dated 5th October 2001.
8   Refer to the article of Suman Sahai. *Can India Handle GM Technology?* Retrieved June 10, 2012, from www.genecampaign.org/Publication/Article/GMtech/ CanIndiaHandleGMTechnology.pdf
9   For details refer *GMO Regulations in India and Their Weaknesses*, p. 3. Retrieved the "*Gene Campaign*" website from www.genecampaign.org/Publication/Article/ gmo-reg-india-weakness-p2=ID1.htm
10 Ibid.
11 Refer to the article of Ashok B. Sharma (2010). *India Needs Biosafety Authority, Not Biotech Authority.* Retrieved May 30, 2012, from www.d-sector.org/article-det.asp?id=1149
12 See *Cartagena Protocol on Biosafety to the Convention on Biological Diversity.* Text and Annexes, Retrieved June 27, 2012, from www.cbd.int/doc/legal/ cartagena-protocol-en.pdf
13 Refer to *A Comment on the: Biotechnology Regulatory Authority of India (BRAI) Bill.* Centre for Indian Political Research and Analysis (CIPRA). Retrieved May 12, 2012, from www.cipra.in/paper/BRAI.html
14 *A Comment on the: Biotechnology Regulatory Authority of India (BRAI) Bill.* Centre for Indian Political Research and Analysis (CIPRA). Retrieved May 12, 2012, from www.cipra.in/paper/BRAI.html
15 See the report of Avik Roy. (2012, March 5). "Beware the Biotech Bill", *The Pioneer*, Retrieved June 30, 2012, from www.dailypioneer.com/columnists/ item/51183-beware-the-biotech-bill.html
16 See *The Biotechnology Regulatory Authority of India (BRAI) Bill 2011- The Bill to End the Right to Safe Food!* At the Greenpeace website. Retrieved June 30, 2012, from www.greenpeace.org/india/Global/india/report/brai%20%20 critique.pdf
17 *Regulatory Regime for Genetically Modified Foods: The Way Ahead.* Indian Council of Medical Research, New Delhi – April 2004. Retrieved July 2, 2012, from http://icmr.nic.in/reg_regimen.pdf

3 The first GM crop of India

The existing controversies around Bt cotton cultivation

Introduction

This chapter deals with the politics of GM technology in India. Currently, the debate on this subject has turned controversial, with different policy coalitions supporting and opposing the technology. This has brought to the fore conflicts in the state structures. The conflict among the proponents and detractors of GM crops was exacerbated in 2010 when the then Environment Minister Mr Jairam Ramesh imposed a moratorium on the release of transgenic brinjal hybrid called Bt brinjal developed by Mahyco, a subsidiary of the global seed company Monsanto[1] despite approval from the GEAC. The minister's decision came after holding public consultations in seven cities, which were attended by approximately 8,000 people. They were organised after widespread protests against the GEAC's recommendation of approval of Bt brinjal in October 2009.

Given the seriousness of the implications, the Indian parliament set up a Parliamentary Standing Committee and the Supreme Court of India the Technical Expert Committee (TEC) to make recommendations on the issue. Both committees submitted their report recommending a moratorium on field trials of GM crops until a mechanism for independent assessment and robust regulatory process has evolved to consider all biodiversity-, health- and consumer rights–related issues. However, the recommendations of both committees did not go well with the PMO as well as the Agriculture Ministry under Minister Mr Sharad Pawar. The PMO and the Agriculture Ministry wanted the Genetic Engineering Appraisal Committee (GEAC) which is under the supervision of the Ministry of Environment and Forests (MoEF), to approve field trials of various GM food crops.

However, the then Minister of Environment and Forests Mrs Jayanthi Natarajan, who was brought to the office after replacing Mr Jairam Ramesh, took the position that no clearances should be given by her ministry's statutory GEAC for field trial proposals until the court makes the final call.[2] Her critical stand of the PM over GM crops likely led to her removal from the ministry. Natarajan was replaced by Mr M. Veerappa Moily. After taking charge of the MoEF, Mr Moily overruled his predecessor and extended the validity of approvals for field trials of GM food crops a couple of weeks later. The new environment minister argued that the clearances had been pending for too long and he cleared them because they had not been banned by the apex court.[3] He also instructed the environment secretary to sit with the Cabinet secretary and other officials to take a united government stand before the Supreme Court. Therefore, all these steps meant agreeing to the views of the PMO and the then Agriculture Minister Mr Sharad Pawar, under UPA-II. The Agriculture Ministry under the UPA-II government then filed an affidavit on behalf of the Union of India opposing the interim report of the TEC, which had recommended a moratorium on GM field trials. Thus the UPA-II government finally managed to grant approval to some GM food crops for field trials just before its term was ending.

Under the new National Democratic Alliance (NDA) government, the GEAC approved field trials for 21 new varieties of genetically modified crops, including staples such as rice and wheat.[4] This move of the NDA government led to dissensions within the Bhartiya Janata Party (BJP), which is the dominating party of the NDA. The Swadeshi Jagaran Manch (SJM)[5] and the Bhartiya Kisan Sangh (BKS),[6] which are important organs of Rashtriya Swayamsevak Sangh (RSS), opposed the government's move to approve field trials. Maharashtra was the first state to grant a No Objection Certificate (NOC) for field trials of five GM crops in year 2015: brinjal, maize, rice, chickpea and cotton. Other states that have issued NOCs for field trials of some biotech crops are Punjab, Haryana, Delhi and Andhra Pradesh.[7] However, apart from these few states, others have opposed the field trials.

It can be observed that apart from the stakeholders within the government, there is no consensus over the extension of the use of GM technology in agriculture. In India, Bt cotton is the only GM crop that has been approved for commercial cultivation by the government regulatory bodies. A decade of experience in Bt cotton cultivation in India has not helped in resolving the conflict over GMOs; rather, it has further intensified the debate among supporters and detractors. The GM technology, heralded in India by Bt cotton, was adopted with the hope that it would help farmers in fulfilling their aspirations by increasing their income through higher agricultural yields. Though Bt cotton helped farmers in increasing their yield, it also raised several other concerns, intensifying the GM technology debate.

Cotton is considered to be one of the most important cash crops in India, and it plays a significant role in contributing to the growth of the national economy. Agricultural scientists have estimated that India has the world's largest area under cotton cultivation, i.e. it represents 20% to 25% of the total global area. It ranks third in terms of cotton production and is behind only China and the United States. Despite this, the yield per hectare of cotton in India has been among the lowest in the world (Gruere, Mehta-Bhatt, & Sengupta, October 2008). Table 3.1 makes the argument clearer.

From this table it is clear that India's average cotton yield has been very low. To deal with this problem, the Indian government, along with agricultural scientists, looked for some technological solutions and found an answer in biotechnology. In 2002, the Indian regulatory body GEAC approved the first GM crop, viz. Bt cotton, for commercial cultivation. This gave rise to a fierce political debate between proponents and opponents of the use of GM technology in agriculture.

Figure 3.1 shows the present top ten countries in the world producing cotton. Figure 3.2 shows the increment in the cotton yield from 1970–1971

Table 3.1 Average cotton lint production, area and yields in the ten leading cotton-producing countries, 1997–2006.

Country	Production		Area		Average Yield	
	Million tonnes	Share (in %)	Million ha	Share (in %)	kg/ha	Rank
China	5.12	24.8	4.7	14.1	1087	6
USA	4.15	20	5.24	15.7	789	14
India	2.27	11	8.65	25.9	273	70
Pakistan	1.89	9.1	3	9	626	23
Uzbekistan	1.08	5.2	1.47	4.4	735	17
Turkey	0.89	4.3	0.66	2	1354	3
Brazil	0.72	3.5	0.9	2.7	832	13
Australia	0.62	3	0.39	1.2	1655	1
Greece	0.39	1.9	0.39	1.2	1002	8
Syria	0.32	1.5	0.24	0.7	1332	4
World	24.84	100	35	100	501	N.A

Source: IFPRI Discussion Paper 00808, October 2008.

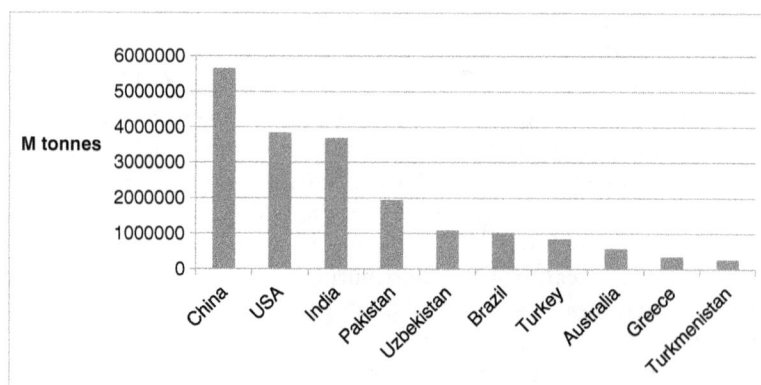

Figure 3.1 Production of cotton lint: top ten producers, average 1994–2018, FAO
Source: Author's compilation using FAO (2018) data.

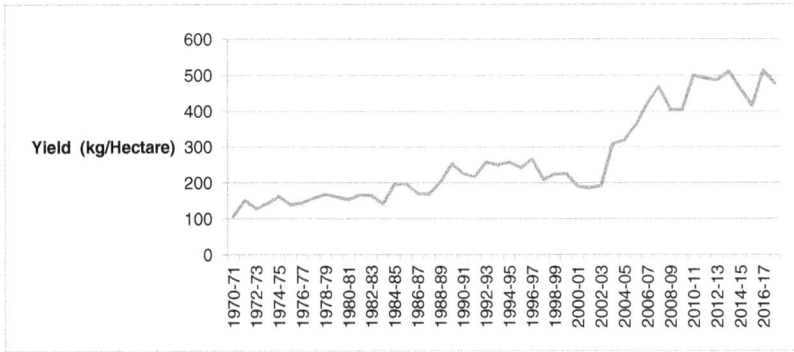

Figure 3.2 Cotton yield trend in India (1970–1971 to 2017–2018)
Source: Author's compilation using data from Agricultural Statistics at a Glance 2018.

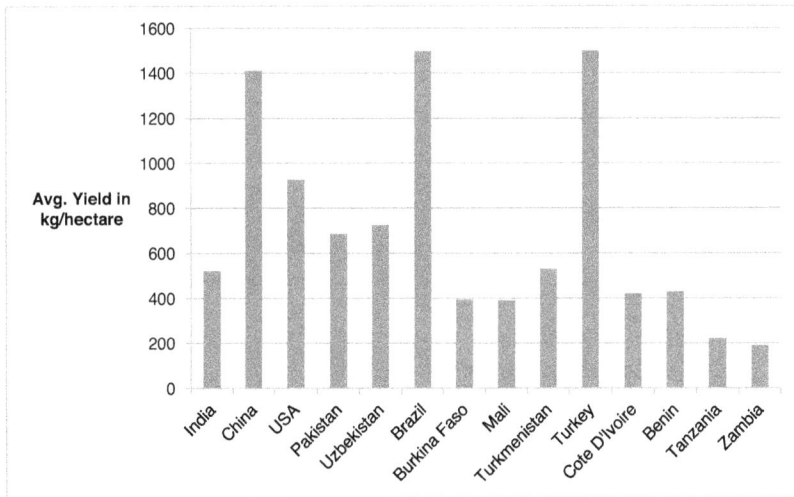

Figure 3.3 Comparative cotton yield/hectare in major cotton-producing countries
Source: Author's compilation using CAI (2020) data.

to 2017–2018 at the all-India level. From the graph it can be observed that there is a steep rise in cotton production from 2002 to 2018. But still if we compare the average yield per hectare of cotton in India with other countries in the world, we find that the average yield per hectare of cotton in India is quite low (see Figure 3.3).

Indian history of cotton farming

Cotton cultivation has a deep history in the Indian subcontinent. According to Glenn Davis Stone, the earliest known cotton in India was from the Indus Valley, and for centuries Indian calicos and muslins were widely traded as luxury goods (Stone, 2004). These calicos and muslins all used to be woven from the yarn of the indigenous species *Gossypium arboreum*, which had short staples (when you think of a cotton plant, a little white puffball probably comes to mind; that is called the 'boll' and each boll contains nearly 250,000 individual cotton fibres or staples; cotton plants are classified on the basis of long and small fibres or staples). The indigenous species had low water requirements and was resistant to indigenous pests. In the late nineteenth century, the British replaced the indigenous or desi cotton with a New World species called *G. hirsutum*. This new species, because of its long staples, was well suited for the British textile mills to produce higher yields. But it was highly vulnerable to Indian pests, and so the adoption failed until pesticide sprays became available in the mid-twentieth century.

In an article on the subject, Supriya Pal mentions that India has been recognised as the cradle of the cotton industry for over 3,000 years (1500 BC to 1700 AD). According to him, Indian farmers have been historically growing plants that have produced cotton for some of the finest and most beautiful cotton fabrics since time immemorial. He further mentions in his paper that the textile industry has been the second largest provider of employment after agriculture. In 2008, the textile industry contributed about 14 per cent of industrial production, 4 per cent of GDP and provided direct employment to over 33 million people (Pal, March 2010). Cotton has become a basic requirement of the textile industry. According to Pal, around 6 million farmers engage in cotton cultivation to meet the needs and consumption of textile industries. These are some of the main reasons why certain corporate industries are in nexus with the state government's campaign for technological solutions to increase cotton productivity.

In India, Bt cotton became the first GM crop to enter the agricultural market and fields. The Indian regulatory bodies RCGM and GEAC approved three Bt cotton hybrids in early 2002: MECH 12, MECH 162 and MECH 184. They all were developed by the Maharashtra Hybrid Seed Company (Mahyco). Mahyco had a licensing agreement with Monsanto to backcross the Cry1AC Bt gene with the conventional cotton breeds (Qaim, Subramanian, Naik, & Zilberman, 2006). According to Qaim, Subramanian, Naik and Zilberman, during the first year of commercial adoption in India, GM hybrids were grown on about 90,000 acres. In 2003–2004, almost 250,000 acres of land were used for Bt cotton cultivation. In 2004, a fourth Bt cotton hybrid was developed by Rasi Seeds and was subsequently commercially approved for cultivation. The area used for its cultivation was estimated to be around 1.3 million acres, which is equivalent to about 7 per cent of the total national cotton area (Qaim, Subramanian, Naik, & Zilberman, 2006). In addition, there is a sizeable black market for unapproved Bt cotton seeds. In 2019, farmers in Maharashtra planted the banned herbicide-tolerant Bt (HTBt) cotton to show their support for the adoption of genetically modified crops. This was an act that could have invited a Rs 1 lakh fine and five years in jail. The farmers' group said that they were protesting the government's cap on this technology, adding they did not want subsidies or loan waivers, just their independence to make a choice.[8] While Bt cotton, a pest-resistant variety that produces an insecticide to combat bollworm, is approved in India, the use of HTBt cotton is banned in India.

Arguments from different sections over Bt cotton cultivation

Monsanto, in its "Sustainability and Corporate Responsibility Report - 2010", says that while demand for food, clothes and houses is growing at a fast pace at present, their supplies from basic resources are not growing significantly at the required pace, and in fact tend to be stagnant. Today farmers need to get more from every acre of land, every drop of water and every unit of energy. The report argues that to bring sustainable agriculture to all parts of the world is the prime motive of the Monsanto team and that they are committed to developing such technologies that can enable farmers to produce more while at the same time conserve more of the natural resources. The report states that Monsanto is going to fulfil its commitment by introducing certain seeds and traits developed through using the applications of biotechnology in the agricultural sector. Monsanto has contended that it would keep its commitments to sustainable agriculture by setting certain goals and endeavouring to achieve them by 2030.[9] These goals are:

- Producing more: Developing improved seeds that would help farmers to double their yields, especially for agricultural crops like corn, soybeans, cotton and spring-planted canola. Monsanto has taken up a $10 million project of the Beachell Borlaug International Scholars Programme, which is committed to improving wheat and rice yields.
- Conserving more: Looking forward to conserving more resources by developing certain improved seed varieties that would use one-third less key resources per unit of output to grow into crops. This would help reduce habitat losses of wild species and improve water quality.
- Improving lives: Help to improve the lives of poor farmers and the people of third world countries who are extremely dependent on agriculture for their livelihood.

The 2010 report from Monsanto states that it is farmers with smallholdings who produce most of the food in developing countries. The conditions of these farmers are generally poorer than the rest of the country's population, and they tend to have less food security than the urban poor. The report further says that these farmers struggle, as they have only limited access to technologies and advanced tools like quality seeds, fertilisers and crop protection products that can enable them to improve their farming conditions.

In this regard, agricultural biotechnology has played a significant role in improving farmers' income. According to the report, in 2008 alone, biotechnology enhanced farmers' incomes by $9 billion, and half of this income gain occurred in developing countries.[10] Today India has become one of the world's leading adopters of new cotton technologies. Monsanto's annual report states that because of the new Bollgard II technology,[11] cotton yields have increased, thereby improving the incomes of some Indian farmers and enabling them to purchase more hectares of land.[12]

The Asia-Pacific Consortium on Agricultural Biotechnology (APCoAB) and the Asia-Pacific Association of Agricultural Research Institutions (APAARI) have together published a report on "Bt Cotton In India". The report mentions that due to the adoption of Bt cotton, the Indian cotton production scenario has changed dramatically. Bt cotton has helped farmers increase their cotton productivity and enabled India to become one of the major countries in the world exporting cotton to other countries. The report further points out that the area under Bt cotton reached 7.6 million hectares in 2008–2009, thus constituting nearly 81 per cent of the total cotton area in India (Karihaloo & Kumar, 2009). As a result, the net cotton productivity reached a record 4.9 million tonnes. According to the latest data of Agricultural Statistics at a Glance – 2018, in 2017–2018, the area under cotton cultivation reached 12.43 million hectares and the overall production of cotton reached 34.89 million tonnes (Directorate of Economics and Statistics, 2018).

However, critics like Suman Sahai and Shakeelur Rahman immediately after approval of Bt cotton in 2002 wrote an article for the *Economic and Political Weekly* on the "Performance of Bt cotton: Data from first commercial crop". The article is based on the results of the small field study conducted in the selected locations of Maharashtra and Andhra Pradesh, two of the six states that have been granted permission to commercially cultivate Bt cotton. In their field study, the authors showed a contrasting picture of Bt cotton's effect in India. According to them, this particular technology has brought no good to Indian farmers (Sahai & Rahman, 2003). Their survey has revealed some of the eye-opening facts. In their study, they mainly compared Bt 162 and Bt 184 belonging to Mahyco-Monsanto with those of local non-Bt cotton varieties such as 'Brahma' and 'Banny'. According to them, Bt cotton was found to be a shorter-duration crop, i.e. 90 to 100 days, while non-Bt cotton's life was found to be of longer duration, i.e. 100 to 120 days. Bt cotton showed comparatively less vigorous growth, with fewer branches and smaller leaves. Its major problem was its premature dropping of bolls. On the other hand, the number of bolls in the non-Bt variety was found to be much higher. The fibre length of the non-Bt variety was comparatively longer and better than its Bt counterpart. One of the most significant findings of their study was that the current variety of Bt cotton in India does not offer any protection against the pink bollworm (*Pectinophora gossypiella*).[13] According to Sahai and Rahman, pink bollworm attacks in the cotton-growing area were found to be severe only after 60 to 70 days. This might be because of two possible reasons. First, the period of expression of the Bt endotoxin might not coincide with the time of the bollworm attack. This would mean that during the time a pest like pink bollworm attacks, Bt cotton would be unable to express the endotoxin gene and therefore unable to offer any protection against that pest (Sahai & Rahman, 2003). The second reason, according to Sahai and Rahman, is that the pink bollworm might not be susceptible to the Bt endotoxin.

G. Shourie David and Y.V.S.T. Sai, in another article, have pointed out that some environmentalists in India opposing Bt cotton cultivation have even gone to the extent of filing Public Interest Litigation (PIL) petitions against the GEAC, the national regulatory body that granted approval to Bt cotton cultivation (David & Sai, 2002). However, scientists like Norman Borlaug, a Nobel Laureate who is known as the father of the Green Revolution, have expressed the need for adopting transgenic crops in a country like India. Even the country's autonomous apex organisation, viz. Indian Council of Agricultural Research (ICAR), recently certified that Bt cotton cultivation on Indian farms is safe as per the test trials that were conducted to understand its efficacy in relation to both the environment and agricultural productivity. David and Sai have pointed out in their paper that a rigorous discussion has been going on between the government, media, researchers and experts to know each other's view on the issue. However, until now there has never been an adequate systematic effort to ascertain the opinions of the farmers on the matter, who are the major stakeholders in agriculture. Keeping this in mind, the authors prepared a survey report where they tried to gauge the reactions and opinions of farmers of Warangal and Khammam districts in the Telangana region of Andhra Pradesh on Bt cotton cultivation. According to them, in both regions farmers have been cultivating cotton, and in the past few years several cotton farmers have committed suicide. According to David and Sai, Bt cotton cultivation in the region is not widely spread but dispersed, and only three or four farmers per village were selected and given Bt cotton seeds by some private or government agents for cultivation over an area of 1 or 2 acres each. Their survey revealed that around 30 per cent of the farmers are ready to cultivate cotton only if they received a suitable price in the market (David & Sai, 2002). Based on their survey report, David and Sai have stated that none of the farmers opposed Bt cotton cultivation on technical grounds. This shows that the concerns of farmers are in direct opposition to those of environmentalists. Their argument is in direct contradiction to the observations made by Sahai in 2002. In her article "The Economics of Bt Cotton", Sahai states:

> Faced with defiant farmers who do not see the logic of 'wasting' 20 per cent of their land, the government is now finding it difficult to convince farmers that this fantastic technology they were promoting all along, does indeed have a downside. Scientists and agriculture departments are already admitting that they have a problem on their hands since

the farmers do not intend to follow any instructions about demarcating insect refuges.[14]

David and Sai (2002) have further argued, based on their survey report, that there has not been much reduction in pesticide expenditure because farmers were unable to distinguish between Bt and non-Bt at the time of pesticide spraying. The farmers' confusion arose mainly because Bt cotton is grown over a minor fraction of the total area under cultivation, and farmers used to spray the same amount of pesticide on all the fields irrespective of whether Bt cotton or non-Bt cotton is grown.

Further, it has been observed that private companies like Mahyco-Monsanto have been lobbying the government regulatory bodies to influence farmers by highlighting short-term goals. For instance, they tell the farmers that if they buy Bt cotton seeds, their productivity will increase and they would make a good profit. We have already observed in the survey report of David and Sai that farmers are not much bothered about the long-term consequences of Bt cotton planting. Their prime concern is to make a profit, and for that they can easily be persuaded by some corporate or government agents.

Glenn Davis Stone in his article has shown how pro-industry agricultural leaders and proponents of Bt cotton cultivation insist that farmers should be given the choice to select a technology according to their own wisdom. For instance, the Andhra Pradesh agriculture minister states:

> Let the farmers finally decide on the usefulness of Bt cotton. Farmers are wise enough to adopt anything good and discard things that do not work.
>
> (Stone, 2007)

However, Stone is critical of this argument and states that it ignores several social aspects, in particular agricultural decision-making. He further states that the information that we directly get from the environment is known as "Payoff Information". The Payoff Information is initially raw in itself and requires further interpretation through discussions and sharing among the community members. Another important aspect that has to be comprehended, according to Stone, is that every human individual should try to adapt according to their environment as well as their social groupings. The external world for any individual is a bundle of contradictions, and among this bundle of contradictions, the individual is capable of acquiring only certain selective ideas, beliefs and behaviours. According to Stone, the method of acquiring these beliefs, ideas and behaviours is simply based on 'rules of thumb', which further gives rise to several biases (Stone, 2004). Some of these biases are:

> *Direct bias*: This is related to a person's social learning psychology and is sometimes unrelated to the person's performance.

Prestige bias: This kind of bias is associated with some successful actors. For instance, people love to follow only those who have achieved some immense success in any particular field.

Frequency-dependent bias: This refers to the unique tendency of adopting those traits which have already been widely adopted by others.

These analyses of Stone delineate how these transmission biases nullify the arguments of pro-industry leaders and scholars that "farmers would adopt any technology or skill or idea or behaviour only and only if it benefits them". According to Stone, when Warangal farmers were asked during the survey about what influenced them to go for Bt cotton cultivation, their reply was not the ubiquitous 'good yield'; their answer was that since other farmers were growing them, they also followed (Stone, 2007). Private companies often strategically donate Bt seeds to some rich farmers, and when these farmers cultivate them in their farms, the growth of cotton crops are usually good during the first season. Therefore, for publicising Bt seeds among the rest of the farmers, the private companies would bus some farmers of a particular village to inspect the field and entice them by showing the spread of healthy crops. Farmers, being naïve and unaware of the complex interaction of nature, soil and microorganisms, are easily persuaded and fall into the trap of the private companies.

According to the article "Biopiracy, GM Seeds and Rural India" written by Priya Kumar, the 'success' of Bt cotton is simply a fabrication created by Mahyco-Monsanto. It is unrealistic to assume that such conglomerates would publish data that would undermine their interests in financial gains through seed monopolisations. According to Kumar, a 2004 report of Monsanto predictably claimed Bt cotton to considerably improve cotton farmers' crop yield returns. The report was supposedly prepared after a deep countrywide study. It claimed that cotton yields increased by 58 per cent, resulting in an increase in farmer incomes by 60 per cent (Kumar, 2009).[15]

However, according to Kumar, just two years earlier, i.e. in 2002, the Research Foundation for Science, Technology and Ecology (RFSTE) undertook a survey study in an attempt to highlight the real effects of Bt cotton on yields. In the survey, RFSTE discovered that in reality bollworm pests attacked Bt crops far more often than they did the simple hybrid and organic cotton crops. According to their research, the claim of 3,300 pounds of yield per acre was never realised, with the highest yield being only 880 pounds per acre (Kumar, 2009). The RFSTE survey report finally concluded that both organic and hybrid cotton-producing rural communities produced an average yield of 1,000 pounds per acre. However, the facts produced by Kumar in her study are extremely controversial and biased.

On 28 August 2011, the *Times of India* (TOI) newspaper claimed that farmers in Bhambraja and Antargaon villages earned huge profits by cultivating Bt cotton in their farms. Both these villages fall within the district of Yavantmal in Maharashtra. According to the TOI news, Yavantmal district is known for its rate of suicide among farmers. The survey report says that among farmers who commit suicide in this district, most are cotton growers. Bhambraja and Antargaon are the only two villages in the district where no suicides have taken place recently and people are prospering on agriculture.[16] According to the TOI news, it is the cultivation of Bollgard or Bt cotton by the farmers of these villages that led to a social and economic transformation. According to the report, farmers of these villages claim that their income from Bt cotton cultivation has not only helped them to get rid of the compounding loans from moneylenders but has also fulfilled all their aspirations of sending their children to the nearest convent schools and getting their daughters married as lavishly as people in cities do. Because of the incomes gained by Bt cotton cultivation, some of the farmers of these villages also managed to buy more acres of land. For instance, a farmer named Raut of Bhambraja village, according to the TOI news, previously had 6 acres of land. But from the increased income through Bt cotton cultivation, he not only managed to buy 2 more acres but also built a pukka house on it. The TOI news further describes that Raut's (farmer) yield from cotton increased from just 6 quintals per acre from conventional or non-Bt cotton in 2001 to about 20 quintals per acre with Bolgard II. Therefore, he now earned Rs 20,000 more per acre due to savings in pesticide.[17]

An article by P. Sainath in *The Hindu* newspaper severely criticised this news from TOI. According to Sainath, TOI published the same news word for word twice in three years, the first time as news and the second time as an advertisement. Sainath states that the piece of information published in TOI is shocking and disheartening. This is because the villagers whom TOI described as prospering and happy with Bt cotton cultivation had a different story to tell the Parliamentary Standing Committee on Agriculture headed by veteran parliamentarian Basudeb Acharia in March 2012.[18] They informed the committee that up to that point, 14 suicides had taken place in their villages and mostly after the coming of Bt cotton. While *The Hindu* was able to verify that nine of the suicides had occurred between 2003 and 2009, the activist groups counted five more. The most shocking part of the story is that all the suicides in the two villages, viz. Bhambraja and Antargaon, took place after 2002, whereas TOI drew a completely different picture of no suicides. The villagers further stated that many plots of land are lying fallow because several farmers have lost faith in farming and have started migrating towards cities in search of some other work. Sainath sees some conspiracy and high-level politics behind the publication of the wrong news of Bt cotton yields in these villages at a time when a parliamentary session was going on to discuss whether to pass the new bill on the Biotechnology Regulatory Authority of India (BRAI). He alleges that the glowing photographs that accompanied the TOI coverage regarding the Bt miracle issue were not taken in Bhambraja or Antargaon. He also alleges that TOI provided wrong information about the villager named Nandkishore Raut. TOI reported that he started earning Rs 20,000 more per acre due to pesticide savings. But the fact of the matter was that when Raut was asked about his experience of Bt cotton cultivation, he replied that Bt cotton was not at all suitable for his village land, as it does not have proper irrigation facilities and fully depends on rain, and Bt seeds are not suitable for rain-fed areas.[19]

Dinesh C. Sharma in *India Today* has reported that the ongoing debate on biotechnology crops in India took a new turn when the American seed firm Monsanto accepted the fact that the pink bollworm pest has developed resistance to Bt cotton in Gujarat. Monsanto further stated that the resistance was natural and expected. The company blamed pink bollworm resistance on the early use of unapproved Bt cotton seeds by farmers and also their refusal to do refuge planting.[20] To tackle the problem of pink bollworm attacks on plants, Monsanto introduced second-generation Bollgard II in 2006, which contained two proteins: Cry1AC and Cry2Ab.

However, this revelation has not surprised the environment action groups. According to them, this is the pattern Monsanto has been following everywhere. Once Bollgard 1 fails, they start pushing Bollgard 2 and recommend that farmers apply more pesticides. Sharma argues that Bt cotton has not only been rendered ineffective but has also led to the detection of some new pests never before reported in India. According to Sharma, Bt cotton is toxic only to the bollworm and does not control any other pests of cotton. Therefore, new sucking pests, which were earlier insignificant, have emerged as major threats and have started causing significant economic losses. Sharma also mentions that cotton productivity has fallen from 560 kg lint per hectare in 2007 to 512 kg in 2009. The pesticide expenditure has also gone up from Rs 597 crore in 2002 to Rs 791 crore in 2009.[21]

Vandana Shiva, in her article "From Seeds of Suicide to Seeds of Hope", argues that the shift from saved seeds to a corporate monopoly on seed supply represents a shift from biodiversity to monoculture in agriculture. According to her, in 1998 the World Bank made some structural adjustment in its policies and forced India to open up its seed sector to global corporations like Cargill, Monsanto and Syngenta. In this way, farm seeds in India started getting replaced by corporate seeds, which need fertilisers and pesticides and cannot be saved. Corporations also managed to prevent seed savings through patent rights and by engineering the seed with non-renewable traits.[22] As a result, poor peasants are compelled to buy new seeds for every planting season. Shiva argues that what was traditionally a free resource for farmers has now become a commodity. The new expense has led to an increase in poverty and indebtedness in India. According to Shiva, monocultures and uniformity increase the risk of crop failure. This is because diverse seeds adapted to diverse ecosystems are replaced by the rushed introduction of uniform and often untested seeds into the market. Shiva further reports that when Monsanto first introduced Bt cotton in 2002, Indian farmers suffered a loss of 1 billion rupees due to crop failure. The company had promised a yield increase of 1,500 kilos per acre. Instead, the yield was as low as 200 kilos per acre. Instead of gaining a net income of Rs 10,000 per acre, farmers ran into losses of Rs 6,400 per acre.[23]

Another major problem that the Indian farmers are facing according to Shiva is the dramatic fall in prices of farm produce as a result of the WTO's free trade policies, which are discriminatory in the sense that they allow wealthy countries to increase their agribusiness subsidies while preventing other countries from protecting their farmers from artificially cheap imported produce. According to Shiva, the $400 billion in subsidies combined with the forced removal of import restrictions is a ready-made recipe for farmer suicides.

In India, the region where the highest number of farmer suicides takes place is Vidharbha in Maharashtra. It is reported that in Vidarbha every year around 4,000 farmers commit suicide, which works out to 10 per day.[24] The region is also known for the highest level of Monsanto's GMO Bt cotton cultivation. According to Shiva, Monsanto's GM seeds create a suicide economy by transforming seeds from a renewable resource into a non-renewable input. Farmers are left with no option but to buy them at high prices from the seed market every year. Shiva makes a comparative analysis of the cost benefits between conventional and Bt cotton seeds. Indigenous cotton varieties can be intercropped with other food crops, while Bt cotton can only be grown as monoculture. Indigenous cotton can grow easily in rain-fed areas, while Bt cotton needs regular irrigation. Further, indigenous varieties have been found to be pest resistant, while Bt cotton, although promoted as bollworm resistant, is prone to several new pests, and to control them, farmers are using 13 times more pesticides than they were using before the introduction of Bt cotton. Shiva argues that Monsanto has been involved in making several fraudulent claims of high yields around 1,500 kg per year. In reality, farmers have only been able to harvest around 300 to 400 kg per year.[25] She further argues that while Monsanto pushes the costs of cultivation up, the agribusiness subsidies in a few developed countries drive down the prices that farmers of developing countries get for their produce. All these factors lead to the creation of a debt trap and a suicide economy for any country whose major share of the population is dependent upon agriculture for livelihood.

A study conducted by the RFSTE shows that due to falling farm prices, Indian peasants have started losing $26 billion annually.[26] Poor peasants are unable to bear this burden and therefore are unable to get rid of their debts. As a result, they are often compelled to sell a kidney or commit suicide. According to Shiva, while seed saving sustains farmers, seed monopolies rob them of their life. The latest biotechnology in the form of GM crops thus helps some private companies to create a monopoly over the regenerative seeds of the earth.

P. Chengal Reddy, who is the president of Federation of Farmers Associations (FFA) in Andhra Pradesh (AP), has criticised Vandana Shiva's point of view regarding Bt cotton cultivation in Indian farms. Reddy has expressed his distress at the growing number of environmental NGOs in India which preach through their private publications and popular media about what Indian farmers should and should not do. He alleges that these environmental NGOs are composed of non-agriculturist members. According to him, they do not know anything about Indian agriculture, and their talks and written articles rarely relate to the realities of Indian farms. He severely criticises one of the articles of Shiva on Bt cotton cultivation in Gujarat. He argues that although Shiva speaks a lot on the plight of monarch butterflies[27] and ladybirds[28] due to Bt cotton cultivation in distant countries, her article does not mention any interviews or talks with the Bt cotton growers of Gujarat. Reddy contends that Shiva is ignorant of certain home truths.[29] According to him, Shiva sounds comical when she says, "In the absence of biosafety capacity building, commercial introduction of Genetically Modified Organisms (GMOs) amount to bio-terrorism". Reddy alleges that some environmental activists like Shiva simply want to promote red-tapism to protect their interests, and in the process tend to block the agricultural progress of the country. According to him, people who are dependent on agriculture for their livelihood want more and more de-controls, but environmentalists propose more and more control on agriculture. He is also critical of Shiva's concern for Indian biodiversity. According to him, biodiversity and food production need not, and in fact do not, go together. Reddy states that:

> Had our farmers continued with 'desi breeds' in the name of preserving 'biodiversity', India would still be importing milk and milk products from other countries and food grains, eggs and chicken would continue to remain beyond the reach of many in lower income groups.[30]

Reddy argues that the job of the farming community is to produce more food grains and other allied items to meet the requirements of the rapidly growing population. He also challenges the claim of Shiva that organic cultivation is better than GM crop cultivation in increasing farm yields.

According to Lianchawii, more than 50 per cent of the insecticides consumed in Indian agriculture go for cotton crops alone. According to her, pesticides cause huge losses to the farmer, as the fields are sprayed 12 to 14 times more in a season in many parts of north India. The American bollworm, considered to be the biggest threat, was responsible for the destruction of about 13 per cent of India's cotton crop production in 2000–2001. The new GM seeds, in the form of Bt cotton, were projected as the solution that can reduce the use of harmful pesticides and make the cotton crop resistant to its natural enemy, the American bollworm. To assess the effectiveness of Bt cotton cultivation, surveys were conducted at different locations funded by various organisations. However, according to Lianchawii, results from the field were varied and inconclusive, as inferences drawn from surveys lacked certainty. The agriculture ministers from AP and Karnataka had already announced the failure of Bt cotton in both states (Lianchawii,

2005). Their statements were supported by some independent organisations which conducted surveys in 2002–2003 in the cotton-growing belts of South India. These organisations were Delhi-based groups such as the RFSTE and the Gene Campaign. Similarly, Greenpeace India also conducted studies in three districts of Karnataka and showed that the average yield of Bt cotton was lower than that of non-Bt cotton. On the other hand, Monsanto kept on proclaiming the success of its three approved varieties of Bt cotton, based on a survey conducted by the company with government officials in five states. The survey reported an increase in yield of 30 per cent and a reduction in pesticide use by 65 to 70 per cent, giving farmers an additional income of Rs 7,000 per acre (Lianchawii, 2005). According to Lianchawii, these differing interpretations only created confusion amongst farmers across the country.

Lianchawii (2005) has mentioned in her article a very important incident regarding illicit Bt cotton cultivation in Gujarat. In late 2001, when cotton farmers in Gujarat were anticipating a bumper harvest, they received an instruction from the government to destroy all their cotton crops by burning them. This order was issued by GEAC, which claimed that it had received evidence that some illicit Bt cotton, unapproved by the committee, was being sown in many parts of Gujarat. Allegedly, these transgenic seeds were sold to farmers by Navbharat Seeds, an Ahmadabad-based company (Lianchawii, 2005). According to Lianchawii, the central authorities requested the Gujarat government to retrieve all illicit Bt cotton that had entered the market. However, the state authorities were reluctant to initiate any action, as it could affect the livelihood of Gujarat farmers. This seemingly incompetent handling of the situation by the state and the GEAC left the markets awash with Navbharat seeds. Lianchawii argues that the decision to destroy the crops punished only the farmers and not the corporation that developed this technology.

According to Jyotika Sood, a study conducted by the Council for Social Development (CSD) on Bt cotton titled "Socio-Economic Impact Assessment of Bt cotton in India" has concluded that the genetically modified cotton has improved farmers' lives with its better yield and higher returns.[31] The study was funded by the farmers' outfit BKS and was released in Delhi. According to this study, 94 per cent of the surveyed farmers were of the view that yields from hybrid Bt cotton seeds were higher than those from non-Bt cotton. Around 87 per cent of the farmers said that their average returns from hybrid Bt cotton seeds were much higher than from non-Bt cotton seeds. The study claims that Bt cotton seeds considerably increased the average returns of farmers by almost 375 per cent to Rs 65,307.82 per hectare.[32]

However, along with these positive implications of Bt cotton cultivation, Sood mentions in her article that the CSD study also highlights some negative consequences that occurred due to Bt cotton planting. According to the study, Bt cotton has increased fertiliser and water usage, resulting in high input costs. The study further indicates that the average consumption of fertilisers has increased from 95 kg per hectare in the pre-Bt cotton period (i.e. 1996 to 2001) to 120 kg per hectare in the post Bt cotton period (i.e. 2002 to 2008). It has shown that farmers who used to irrigate their cotton crops three times in a cycle before Bt was introduced now have to irrigate their crop five times. This shows that Bt cotton is a water-intensive crop. Therefore, the average irrigation costs per hectare also increased from Rs 355 per hectare in the pre-Bt cotton period to Rs 813 in the post-Bt cotton period. The increase is also attributed to higher diesel prices in the country.

According to an article published in the *Hindustan Times* by Zia Haq (2012), India's Bt cotton dream has gone terribly wrong, and an internal advisory of the ministry of the GOI clearly recognised this when it stated for the first time:

> Cotton farmers are in a deep crisis since shifting to Bt cotton. The spate of farmer suicides in 2011–12 has been particularly severe among Bt cotton farmers.[33]

According to Haq, Bt cotton's success lasted merely for five years. Thereafter yields have been falling and pest attacks have been going up. Haq states that the rising costs of pesticides are eroding the returns of farmers and argues that Bt cotton is no longer as profitable as it used to be. The statement of the internal advisory board of the ministry is based on observations from the Indian Council of Agricultural Sciences, which administers farm science, and the Central Cotton Research Institute, the country's top cotton research facility.[34]

However, later reports in the *Hindustan Times* throw doubts on this account of 'deep crisis'. According to the paper, there seemed to be some communication gap within the ministry as India's deputy director-general of crop science, Mr Swapan Kumar Dutta, maintained that he was unaware of the note and reiterated that Bt cotton still continued to drive India's cotton production. The agriculture secretary, Mr Prabeer Kumar Basu, could neither deny nor confirm this information.

According to a survey conducted by Dr Sudhir Kaura, who is a biotechnologist and a geneticist, in many districts of Haryana state in India, sheep are dying after grazing on Bt cotton plants and perish within 10 to 15 days. However, some sheep manage to survive somehow. According to him, many buffaloes also die after grazing on Bt cotton leaves. However, they take a longer time to die when compared to sheep. They also may develop other disorders such as the inability to conceive properly or regularly, or not at all, and may give birth to significantly smaller sized offspring much before the normal delivery time. Other symptoms include protrusion of the uterus, skin allergies, reduction in milk yield, dysfunction of teats, etc.

According to Dr Kaura, many farmers or labourers who handle animal feed containing Bt cotton seeds or oilcakes are experiencing itching and allergy symptoms on the chest, back, neck, hands and even their genitals. The local medical practitioners of the village term this itching scabies, which requires costly medicines for treatment. This causes further distress to the poor labourers in India, whose daily earning is only $1 or 2 and sometimes even less than that.[35]

Analysis of the government-appointed committee reports

The Indian government formed two essential committees to look into the matter of the GM technology controversy. A TEC was set up on the order of the honourable Supreme Court of India on 10 May 2012 following a writ petition filed by Aruna Rodrigues and others (Final Report of the Technical Expert Committee, 2013, p. 1). To deal with the same matter, a Parliamentary Standing Committee on Agriculture was formed in the fifteenth Lok-Sabha. The committee, in its fifty-ninth report in 2013–2014, talked at great length about the controversies surrounding GM crops (Committee on Agriculture-59th Report, 2013–2014).

The TEC was formed to give recommendations on overall matters related to risk assessment of environment and health safety issues, as well as to evaluate specific conditions required for open field trials of GM crops. Before evaluating the conditions for open field trials, the TEC report mentions the need to consider some international agreements related to food safety conservation. An important one is the Codex Alimentarius Commission (CAC), of which India became a member in 1964. The CAC provides guidelines and standards for food safety. It revises its guidance documents from time to time on issues dealing with food safety, including foods derived from biotechnology. India is also one of the important countries that adopted the Rio Declaration on Environment and Development in 1992. According to the TEC report, "the Rio Declaration on Environment and Development is a statement of 27 principles for the purpose of guiding sustainable development across the world" (Final Report of the Technical Expert Committee, 2013, pp. 14–15).

India is also associated with the Convention on Biological Diversity (CBD) along with 193 countries. According to the TEC report, CBD is a legally binding treaty that holds three stated aims (Final Report of the Technical Expert Committee, 2013, p. 14). They are:

1   the conservation of biological diversity
2   the sustainable use of the components of biological diversity
3   the fair and equitable sharing of the benefits arising out of the utilisation of genetic resources.

Besides these, CBD comprises 42 articles outlining ways for the identification, conservation and management of biodiversity, including guidance documents on issues related to the risk assessment of living modified organisms (LMOs).

India has also signed the Cartagena Protocol on Biosafety (CPB) to the CBD (Final Report of the Technical Expert Committee, 2013, pp. 14–15). The CPB is an international agreement aimed at

> ensuring an adequate level of protection for the safe transfer, handling and use of living modified organisms (LMOs) resulting from modern biotechnology experiments and uses which may have adverse effects on the conservation and sustainable use of biological diversity, and also taking into consideration risks to human health, and trans-boundary movements of LMOs across countries.
>
> (Final Report of the Technical Expert Committee, 2013)

According to the TEC report, "the Cartagena Protocol comprises of forty articles covering handling, transfer, risk management, capacity building (in biosafety), public awareness and participation, socio-economic consider-ations and other issues" (Final Report of the Technical Expert Committee, 2013, pp. 14–15).

Therefore, CPB and CAC are the two international-level protocols and commissions that provide the principal guidelines for the biosafety of LMOs, including GM crops. According to the TEC report, "the apex regulatory body for evaluation of GMOs/LMOs is the Genetic Engineering Appraisal Com-mittee (GEAC) located in the Ministry of Environment and Forests (MoEF)" (Final Report of the Technical Expert Committee, 2013, p. 15). "The second arm of the regulatory body is the Review Committee on Genetic Manipula-tion (RCGM) located within the Department of Biotechnology (DBT) of the Ministry of Science and Technology" (Final Report of the Technical Expert Committee, 2013, p. 15). RCGM examines the health safety and molecular characterisations of LMOs/GMOs. Similarly, environmental safety comes under the complete domain of the GEAC. However, the RCGM also exam-ines information on environmental safety. Further, the responsibilities of the RCGM include the "review of applications for research projects involving recombinant DNA and animal experimentation" (Final Report of the Tech-nical Expert Committee, 2013, p. 15). According to the TEC report, both RCGM and GEAC have approximately 30 members each. They include nom-inated representatives of government departments and agencies, researchers and administrators. Further, it has been reported that

> GEAC and RCGM/DBT have produced a number of documents and guidelines covering Recombinant DNA Safety, guidelines for research on transgenic plants, Guidelines and Standard Operating Procedures (SOPs) for confined field trials, guidelines for the safety assessment of foods derived from genetically engineered plants (prepared by Indian Council of Medical Research).
>
> (Final Report of the Technical Expert Committee, 2013, p. 16)

The TEC report says there is also a draft available for the "Guidance for Information/Date Generation and Document for Safety Assessment of Regulated Genetically Engineered Plants", which talks about the studies and tests to be carried out for the safety assessment of GM plants.

Before finalising the report, the TEC submitted an interim report to the Supreme Court on the recommendations being made by TEC members. The interim report was of the view that there are weaknesses in the regulatory mechanisms of the bodies giving approval to field trials. It says that "the practice of allowing the applicant to choose the site for conducting the trials and leaving the onus on the applicant to ensure conditions for safety introduces chances for violation of conditions for safety" (Final Report of the Technical Expert Committee, 2013, p. 16). The TEC had been informed that plots of land were being leased to the applicant or tester for at least three years for conducting field trials, which the TEC did not consider appropriate. According to the TEC, the regulatory authority should have the onus to establish and certify sites for field trials either in ICAR institutes or in state agricultural universities (SAUs). Such sites should have suitable isolation and access restrictions (walled area), with appropriate facilities for conducting field trials, associated biosafety tests and adequate facilities for disposal (i.e. incineration of plant material).

Further, the TEC has suggested in its interim report that post-release monitoring (PRM) is an important aspect in the assessment of environment and health safety issues. According to the report, this aspect has not received adequate importance from the regulatory bodies. The committee felt it was ironic that while the importance of socio-economic considerations, sustainability and development goals have been very well recognised in many of the international agreements which India has signed and is a part of (e.g. CBD and CPB), these considerations have not adequately figured in the regulatory system of the country. So according to TEC, this deficiency needs to be recognised and corrected.

Regarding health safety issues, the TEC has asked the regulatory bodies to look for chronic toxicity rather than acute toxicity, tests for which were done on rodents. This is because, according to the TEC members, food is consumed by humans and animals for their entire lifetime, and nutritional stress can lead to unintended and adverse effects after long-term exposure. Therefore, for food safety testing, acute toxicity testing will not be of much help. Chronic toxicity and intergenerational testing are the primary requirements for safety testing in food. However, the regulatory bodies are less concerned with chronic toxicity tests. Most of the tests that have been done are for acute toxicity. The TEC's interim report also has shown concern about the use of HT technology and recommended that this particular technology might not be suitable in the Indian socio-economic context. This is because this kind of technology is more suitable for large farm sizes of hundreds of acres. In India, the average farmland is small, with an area of about 3.3 acres. Therefore, the TEC has recommended a moratorium on field trials of HT crops until the whole issue is examined by an independent committee.

The TEC has also shown concern over the effect of transgenic crops on the biodiversity of other crops in India once these transgenic crops manage to get approval for open field trials. According to the TEC, crop diversity represents an important cultural heritage, and so special measures should be taken to preserve it. The TEC has recommended collecting information on biosafety prior to field trials and doing experimental tests in this direction before bringing GM crops out of the laboratory.

It has also discussed whether it is possible to evaluate GM crops under greenhouse conditions and whether it is possible to replicate outside agro-ecological conditions in greenhouses. According to the committee, plants have evolved numerous responses and adaptations to changes in the environment. These changes can be in the form of light, water, humidity, temperature, wind, seasons and soil quality. Therefore, according to TEC members, the properties of plants will depend on the complex relationship between these given factors and how these interdependencies affect plant growth, development and response. The growth and development of plants are very sensitive to these factors. Therefore, according to the TEC, in general it is very difficult to replicate in the greenhouse the uncertain conditions that are outside. The TEC therefore recommended that the applicants undertake the studies on a case-by-case basis in consultation with the regulator to do event selection either in the greenhouse or under confined conditions outside the greenhouse. However, it did not suggest any recommendations for examining the feasibility of prescribing validated protocols and active testing for contamination at a level that would preclude any escaped material from causing an adverse effect on the environment.

The TEC has argued that the concerns related to biosafety have gone unnoticed and unaddressed in the course of the regulatory process leading to the approval of open field trials of GM crops. It has further alleged that scrutiny of the biosafety information was done by a committee of the regulatory body which lacked full-time qualified personnel for the purpose. It says that it has come across examples of problematic data that failed to establish the health safety of Bt cotton and Bt brinjal and left unanswered several questions about the overall safety of Bt in food crops. This led the TEC to recommend a ten-year moratorium on Bt in food crops. The issue has been revisited for the purpose of the final report.

All the recommendations mentioned here were given by TEC in its interim report. Following the submission of this interim report, a sixth member was appointed to the TEC. Initially it had only five members. The order for the appointment came from the Supreme Court, which further directed the TEC to submit the final report. For this purpose, the TEC held eight meetings in New Delhi between December 2012 and April 2013, where extensive discussions and exchange of views took place (Final Report of the Technical Expert Committee, 2013, p. 24). The prominent participants were members of the National Academy of Agricultural Sciences, interdepartmental groups comprising secretaries from the Department of Agriculture, senior officials from the MoEF and DBT and senior researchers from ICAR institutes. A separate meeting was held with the secretary of the DBT. The TEC received written submissions from Prof. Deepak Pental, Association of Biotech-led Enterprises (ABLE), National Seed Association of India and others (Final Report of the Technical Expert Committee, 2013, p. 24). All the viewpoints collected through meetings with several organisations and institutes were thoroughly discussed within the TEC. For this purpose, it held meetings at National Institute of Nutrition in Hyderabad from 18 April 2013 to 10 May 2013 in order to give shape to the final report (Final Report of the Technical Expert Committee, 2013, p. 24). In its final report, it repeated most of the points mentioned in the interim report. Based on the deliberations and examinations/study of the safety dossiers, the TEC concluded that there are major gaps in the regulatory system. Therefore, according to TEC, these gaps need to be fixed before approval for open field trials can be granted.

The second committee to assess the prospects and effects of the cultivation of genetically modified food crops was set up by parliament. This Parliamentary Standing Committee on Agriculture was headed by Mr Basudeb Acharia. The committee's final report came in the form of the Fifty-ninth Report on Action Taken by the Government on the Observation/Recommendations contained in the Thirty-seventh Report of the Committee on "Cultivation of Genetically Modified Food Crops – Prospects and Effects Pertaining to the Ministry of Agriculture".[36] In its final report, the committee has argued that it is not satisfied with the replies furnished by the government regarding the recommendations mentioned in the thirty-seventh report. The committee therefore has reiterated its earlier "recommendations and desire that further research and development on transgenics in agricultural crops should be done only in strict containment and field trials should not be undertaken till the government puts in place all regulatory, monitoring, oversight, surveillance and other structures" (Committee on Agriculture-59th Report, 2013–2014, p. 10).

It mentions the allegation made by Dr Pushpa M. Bhargava, who was appointed by the Supreme Court as a member of GEAC.[37] He had pointed out that Bt cotton technology is not a sustainable one. The death of cattle and other livestock in AP after grazing on Bt cotton plants raised serious doubts about the safety of Bt cotton as feed. The committee wanted to know how the regulatory mechanism had missed the 30 per cent increase in toxic alkaloid content in Bt brinjal and approved it for environmental release (Committee on Agriculture-59th Report, 2013–2014, p. 11). This could have adversely affected the environment and human and livestock health.

The department replied in its Action Taken Note that the observation made by Dr Bhargava about the failure of Bt cotton due to the development of insect resistance is incorrect and appears to be based on allegations made by some civil society activists. It has further reported that there are no reports anywhere in the world about the development of resistance by insects to the Bt protein in a cultivated field. It has argued that there are, however, reports available of laboratory experiments that have sought to understand the phenomena of resistance development in the insect. It is in no way scientifically justified to interpret the laboratory observations in the context of field conditions. The department has further suggested that there is adequate scientific evidence to prove that cry proteins found in transgenic crops are not toxic to higher animals such as goats, sheep, etc. Referring to the investigation by the Andhra Pradesh State Department of Agriculture on the death of cattle and sheep due to grazing on Bt cotton fields, the department says that the samples tested in this regard were found to contain high levels of nitrates, nitrites, hydrogen cyanide residues and organophosphates, which were the actual cause of animal deaths. It is believed that they might have come from the soil, fertilisers or pesticides used in cotton cultivation.

However, the committee was not satisfied with this reply from the government and asked why the department remained completely silent on the question of how the regulatory mechanism missed the 30 per cent increase in toxic alkaloid in Bt brinjal and went on to approve it for open field trials.

The committee pointed out that there is some controversy over the exact role played by the GEAC and the MoEF. It says the rule that came into force in 1989 very clearly and unambiguously assigns the role of granting approval for environmental and commercial release to GEAC. However, the committee has observed that

> at some places, the authority of GEAC to accord approvals was truly reflected, but at others it was couched as 'recommendation of GEAC to accord approval' and at still others it was stated that GEAC accorded approval for environmental release and had no role in commercialization of GM crops.
>
> (Committee on Agriculture-59th Report, 2013–2014, p. 16)

The committee has therefore strongly recommended correcting this uncertainty over GEAC's role, as it is not in the interest of the regulatory mechanism, especially in a sensitive matter like GM technology.

The department, in its Action Taken Note, submitted that "as per the Rules 1989, under the Environment Protection Act, 1986, the regulatory powers for environmental release of Genetically Modified Organisms (GMOs) rest only with the GEAC". It has further clarified that the commercial use of any technology is subject to laws, regulations and policies followed by ministries of the central and state governments. Ultimately, the ministries would be responsible for deploying modern technologies in agriculture, health care, process industry, environment protection, etc., which they deem suitable for societal and local needs.

The committee notes that this reply from the government shows that GEAC will have only a regulatory role. This means that it will no longer have the role to approve proposals related to release of GMOs and products in the environment, including approvals for experimental field trials.

The committee has taken into consideration the views of the Food and Agriculture Organization (FAO), World Health Organization (WHO) expert panel and International Assessment of Agricultural Knowledge, Science and Technology for Development (IAASTD) on the use of antibiotic-resistant marker-free gene technology while creating GMOs. The extra care in considering such views is due to the possibility, though remote, of the transfer of genes from GM crops to cells of the body or to the bacteria present in the gastrointestinal tract. In this regard, the GEAC has taken the stand that since the technology for generating marker-free genes is available, it is a matter of policy whether to allow GM crops with antibiotic resistance markers. The committee has expressed its displeasure at the response of the GEAC. According to the committee, the GEAC has shown a complete lack of concern for its role and responsibility and has tended to favour the industry by giving approval for field trials. Therefore, it has recommended that the government not leave such crucial decisions to GEAC and that it should come up with a clear-cut policy in this regard immediately.

The committee has observed that the same groups of people are involved in development of technologies and in the assessment, evaluation and approval of those technologies. Therefore, it has recommended that the government make changes in the composition of the GEAC and other bodies so that the conflicting roles played by some of them can be done away with.

It has discussed the Nagoya-Kuala Lumpur Supplementary Protocol (N-KLSP), which aims for the conservation and sustainable use of biodiversity by laying down international rules and procedures on liability and redressal for damages resulting from LMOs. It has further noted that the government is already going through the process of examining domestic laws on liability and redressal regarding damages from LMOs, as mentioned in Article 2 of the N-KLSP. The committee wants the government to complete this whole process in a reasonable time and fix quickly any loopholes found in light of the N-KLSP.

Most of the recommendations on biodiversity, health and environmental safety made by the Parliamentary Standing Committee on Agriculture have also figured prominently in the TEC. However, the Committee on Agriculture's most important work relates to the field studies it conducted. It had extensive interactions with farmers and has observed that farmers achieved no significant socio-economic benefits due to the introduction of Bt cotton. It further reports that "on the contrary, being a capital intensive agriculture practice, the indebtedness of the farmer had grown massively, thus exposing them to greater risks" (Committee on Agriculture-59th Report, 2013–2014, p. 58). Therefore, according to the committee members, Bt cotton cultivation has only brought miseries to the small and marginal farmers who constitute more than 70 per cent of the farmers in India.

The ministry, in its Action Taken Note, has argued that it would be incorrect to attribute the problems to Bt cotton, as it has effectively controlled bollworms thus preventing yield losses from an estimated damage of 30 to 60 per cent during 2002–2011. According to the government sources, the yields are estimated to have increased at least by 30 per cent due to effective protection from bollworm damage. "All India average yield, which was 189 kg lint per hectare in 2001, increased to 491 kg lint per hectare in 2011" (Committee on Agriculture-59th Report, 2013–2014, pp. 58–59). According to the government sources, about 9,400 million tonnes of insecticides were used for bollworm control in 2001, which declined to only 222 million tonnes in 2011 (Committee on Agriculture-59th Report, 2013–2014, pp. 58–59). Further, the per-hectare income of the farmers increased from Rs 7,058 in 2000 to Rs 16,125 in 2010 under rain-fed conditions and from Rs 15,370 in 2000 to Rs 25,000 in 2010 under irrigated conditions (Committee on Agriculture-59th Report, 2013–2014, p. 59).

Contradicting this government claim, the committee has argued that the farmers with whom it interacted gave a different story. According to the committee, "the first-hand experience gained by the Committee is ample proof to show that the miseries of farmers have compounded since the time they started cultivating Bt cotton" (Committee on Agriculture-59th Report, 2013–2014, p. 59). Therefore, it has asked the government to appreciate the ground reality and not thrust commercial cultivation of Bt cotton on farmers.

Literature reflecting the controversies around GMOs

Herring and Rao (2012) have thrown some light on the existing controversy on GM crops in India. According to them, the allegation that "Bt cotton has failed" has originated from a loose coalition of NGOs often connected to transnational advocacy networks. The NGO narrative explicitly claims that Bt technology has lowered farm yields, its adoption has driven farmers into debt because of high seed prices and agronomic failure has often resulted in catastrophes (Herring & Rao, 2012). However, proponents of transgenic cotton have been arguing that evidence from the field suggests that the use of Bt cotton has been a success.

Herring and Rao have stated that most of the peer-reviewed studies have confirmed the success story of Bt cotton, and they are not aware of any such studies in peer-reviewed journals to show its failure. About the fluctuation in the yield of Bt cotton in different parts of India, Herring and Rao have argued that since India has heterogeneous agro-climatic and socio-economic conditions, different cultivars[38] have been used in different areas in different years. Because of such agronomic differences, cultivars that grow well in one region might not develop so well in another. Farmers would also frequently switch to grow hybrid crops. Now each hybrid crop contains a different germ plasm. Bt technology confers only one trait. So according to Herring and Rao, some hybrids with this trait would do better than others. The only thing that Bt hybrids have in common is this particular trait that makes the plant resistant to the specific insect (Herring, 2006). According to Herring and Rao's estimation, there are around 800 legal hybrids with this specific trait which has been generated through Bt gene insertion. The mechanism for obtaining this trait is exactly the same in all the cultivars. Here Herring and Rao mention the very important point that to measure the effect of Bt technology, it would be ideal to compare two isogenic cultivars, one with and one without the transgene, just to isolate the technological effect. Herring and Rao claim that in this case, cultivars with Bt technology have always shown higher yields. Therefore, if the cultivars are not isogenic, then the result obtained cannot be reliable (Herring & Rao, 2012). So according to Herring and Rao, the reports claiming failure of the technology might have arisen because of a different problem that is difficult to correct. One such problem is the fraudulent practices in an unregulated seed market. According to them, in the field it has been observed that some seed packets are labelled "Bt" but they do not actually contain Bt cotton seeds. They are simply "duplicates" missing the transgene and so they do not produce the cry proteins that provide insect protection (Herring & Rao, 2012). It has been observed that in the Warangal district duplicate seed packets were sold with the name "Mahaco" to trick farmers into thinking that it was "Mahyco" (Herring & Rao, 2012). Therefore, failure to control bollworms on these plants cannot be blamed on Bt technology. According to them, this was not the failure of Bt cotton, but the failure of information in an unregulated seed market. Herring and Rao further report in their article that Warangal and Vidarbha have figured in media reports of failure of Bt cotton without any proper inquiry into the matter. Another important point that needs to be highlighted here is that cotton cultivation usually fails when water is inadequate, irrespective of the fact whether it is Bt or non-Bt cotton. That is why government agencies in India have strongly advised against cotton cultivation in drought-prone marginal areas without proper irrigation facilities. As of now, there is no evidence that the addition of the Bt transgene would affect drought tolerance one way or the other. Drought tolerance is one of the traits considered to be of prime importance by farmers. But to now this particular potential of biotechnology has not been realised.

Haribabu Ejnavarzala (2014) in his article has analysed that the companies that produced the first generation of Bt cotton had anticipated that the bollworm would develop resistance against the Bt toxin, and therefore they are on the lookout for an opportunity to release second-generation Bt technology. The regulatory bodies have also failed to conduct any independent field trials to inquire whether the bollworm has really developed any resistance against the Bt toxin (Ejnavarzala, 2014). According to Ejnavarzala, the regulators simply accept the information provided by the company. He explains that "in the context of GM crops, one should keep in mind that the host and pest co-evolve and the pest develops resistance against the toxin after sometime due to adaptation and mutation". This necessitates further improvement in the crop through genetic engineering to increase its capability to fight against the pest. In this way, a continuous battle is going on between the host and the pest. Ejnavarzala makes the point that perhaps the companies can easily use this as an opportunity to introduce the second generation of the Bt cotton seed (Bollgard II) even when the indications of pest resistance are not visible. In his own words,

> the company that owns the Bt technology employed the strategy of planned obsolescence by introducing the second generation of Bt seed on the basis of the evidence provided by its own crop surveillance staff.
> (Ejnavarzala, 2014)

He argues that it is quite possible that the second generation of Bt cotton seed was introduced for the sole purpose of ever-greening the patent and to increase its price. He reports that in some states where farmers were using the first-generation Bt seeds, they did not find the pink bollworm in the Bt cotton fields. The first-generation Bt cotton seeds were still providing resistance against the bollworm, even though the company deliberately stopped producing them, declaring them to be obsolete, and then introduced the second generation of Bt cotton. Therefore, even if the farmers want to use the first-generation Bt cotton seeds, it is not possible, as they are no longer available in the market (Ejnavarzala, 2014). The only ones available now are the second-generation Bt cotton seeds.

In an article about capitalism, biotechnology and the environment, Friedman argues that the capitalist economy is based on the production of commodities (Friedman, 2015). These commodities are then sold in the market in order to realise profits. According to Friedman, the particular function of a commodity is of secondary importance to its owner. The primary importance of any commodity lies in its capitalisation and realisation as a sum of money. Capitalist growth entails looking for ways to convert the ever-increasing and renewable resources of the natural world into commodities. According to Friedman, in a capitalist world, producers are driven by competition to expand their market shares and scale of production. Realising profits is a matter of economic life and death for them. Therefore, in order to make a profit, they mass-produce cheap, uniform consumer goods by reducing the cost of production (Friedman, 2015). Agriculture and food production are not exceptions to this, as exemplified by today's vast expansion of nutritionally absent fast foods. According to Friedman, the development and deployment of GMOs as food or cash crops is one of the strategies of capitalists to face production challenges. The GM technology, according to Friedman, is meant to address "problems of unit cost, marketability, and production cycle velocity by engineering transgenic organisms that have rapid growth rates, physical attractiveness, novel nutrient combinations, and are pest or herbicide or draught or cold resistant" (Friedman, 2015). Friedman talks about how biodiversity can get affected through homogenised capitalist production. According to him, the dynamics of complex ecosystems run counter to the capitalist producers, who intend to seek absolute control over the production process with the sole aim of eliminating those variables that increase the cost of production and decrease the marketability (Friedman, 2015). GMOs in this regard would appear to be a new tool to enable the capitalists to get a greater degree of control over seeds. According to Friedman, science might have emerged as a social activity, but now it is being increasingly shaped by the dominant institutions, social relations and worldview of our society. For a very long time, elite economic and political interests have set priorities for science. With the emergence of the neoliberal era, science has increasingly come under the influence of private capital. So today, according to Friedman, the priority for biotechnology scientists is to produce technology that can increase profits for their firms (Friedman, 2015). GMO is one such technology that has been developed by them in this regard. It may have helped in increasing the production of small varieties of crops, but that has come at the cost of the destruction or extinction of genetic diversity. Today in India, 90 per cent of the cotton that is grown is the Bt variety. The indigenous variety exists no more, and farmers are compelled to buy Bt cotton seeds from the market since they do not have any other choice.

Conclusion

From this analysis, it is clear that Bt cotton cultivation raises lot of complex questions. After going through the analysis of different survey reports, books and articles, one can argue that its cultivation cannot give higher yields across all regions in India. Bt cotton requires proper irrigation facilities and therefore cannot give higher yields in rain-fed areas. Its cultivation raises several other concerns, such as the preservation of biodiversity. We also have to accept the fact that the Bt cotton seed is a corporate creation and is non-regenerative, artificial and man-made. If its use is going to increase, it would increasingly make farmers' lives dependent on the corporations, as farmers would not be able to save seeds for the next season's cultivation. Therefore, Bt seed technology can help farmers increase their agricultural yields only in certain specific circumstances and in the short run.

Monsanto, which was the first to develop the GM crop, has been strongly promoting its use. In its annual report in 2010, it mentions that with the growing population in the world, the demand for food, clothes and houses has also grown at a great pace. But supplies from natural resources have not matched the pace of the growing demand. Therefore, Monsanto says it is committed to developing such technologies that enable farmers to produce more while at the same time conserving more natural resources. Monsanto claims that its improved GM seed varieties would help farmers conserve resources by increasing the crop yield even in smaller areas. Monsanto's report (2010) mentions that in developing countries like India, it is the farmers with small holdings who produce most of the food, but they are generally poorer than the rest of the population in the country. Monsanto has claimed that farmers' lives have improved with the adoption of its Bt cotton technology. It has stated in its annual report that the new Bollgard II cotton crops have helped farmers increase their yield. The higher yields have increased the net incomes of farmers and enabled them to purchase more land. So in a way, the adoption of Bt cotton seeds has helped Indian farmers improve their economic condition.

However, this claim of Monsanto has been challenged by many civil society groups as well as reports from committees set up by the Indian parliament and the Supreme Court. Sahai, founder of the NGO Gene Campaign, has alleged that the current variety of Bt cotton in India does not offer any protection against the pink bollworm. However, the country's autonomous apex organisation ICAR has certified that Bt cotton on Indian farms is safe as per test trials conducted to understand its efficacy in relation to the environment and agricultural productivity. The TEC and the Parliamentary Standing Committee appointed by the government to give recommendations on GM crops have raised serious concern in their reports regarding biodiversity, health, sustainability, roles of the regulatory bodies and the economic condition of cotton farmers. The committee has recommended that the government reform the composition of regulatory bodies like the GEAC and others, as the same set of people involved in the manufacture of GM technologies is also involved in the assessment, evaluation and approval of those technologies. The committee has rejected the claims of Monsanto, ICAR and other promoters of GM technologies that the economic condition of farmers has improved after their adoption of Bt cotton.

From this analysis and literature review, it can be said that both within and outside the government, there is no unanimous view on GM technology. Some people vociferously support its use, citing several benefits that can be reaped from it. But others oppose its use, citing several concerns over environment, health and ownership rights. The next chapter will talk about a basic theoretical framework that will consider the various concerns of those supporting or opposing GM technology in order to analyse how this affects the policy processes.

Notes

1 See *The Hindu* newspaper dated 9 February 2010 in the national column under the title "Moratorium on Bt Brinjal". Retrieved from the website www.thehindu.com/news/national/Moratorium-on-Bt-brinjal/article16813609.ece

2 See *The Hindu* newspaper dated 1 July 2016 in the national news section. Nitin Sethi is reporting on "Jayanthi Natarajan Opposes Pawar's Views on GM

Crops, Wants Field Trials Put on Hold". Retrieved December 25, 2016, from www.thehindu.com/news/national/jayanthi-natarajan-opposes-pawars-views-on-gm-crops-wants-field-trials-put-on-hold/article4982776.ece

3 See the *Hindustan Times* newspaper dated 28 February 2014 in the India news section. Chetan Chauhan reports the news titled "Veerappa Moily Clears Field Trials of GM Crops". Retrieved December 25, 2016, from www.hindustantimes.com/india/veerappa-moily-clears-field-trials-of-gm-crops/story-Sgps5nLz 9P6AtoFven4Y6H.html

4 See the *Hindustan Times* newspaper dated 15 July 2014 in the India news section. Chetan Chauhan is reporting the news titled "Government Approves Field Trials for Varieties of GM Crops". Retrieved December 28, 2016, from www. hindustantimes.com/india/govt-approves-field-trials-for-varieties-of-gm-crops/story-G4EKDyNBBMWGLM8nU9go3K.html

5 See the *Deccan Herald* newspaper dated 19 June 2014. The news is titled as "RSS Wing Opposes Approval for GM Crop Field Trial". Retrieved December 28, 2016, from www.deccanherald.com/content/420663/archives.php

6 See the DNA daily news and analysis newspaper dated 30 July 2014. Mayank Aggarwal and Amita Shah are reporting the news titled "RSS Wings Oppose GM Crop Trials, Government Says Decision only After Consultations". Retrieved December 28, 2016, from www.dnaindia.com/india/report-rss-wings-oppose-gm-crop-trials-government-says-decision-only-after-consultations-2006487

7 See the *Business Line* newspaper dated 30 January 2015. The news is titled as "Maharashtra Approves Field Trials of Four GM Crops". Retrieved December 28, 2016, from www.thehindubusinessline.com/news/national/maharashtra-approves-field-trials-for-four-gm-crops/article6839527.ece

8 See the news report on "What's the Fuss Over the New Variety of GM Cotton that Farmers Are Batting for" published by *The Print* on 3 July 2019. Retrieved from https://theprint.in/india/whats-the-fuss-over-the-new-variety-of-gm-cotton-that-farmers-are-batting-for/257455/

9 *United in Growth*, Monsanto Company, 2010 - Sustainability and Corporate Responsibility Report. Retrieved July 23, 2012, from www.monsanto.com/SiteCollectionDocuments/2010-csr-report.pdf

10 *United in Growth*, Monsanto Company, 2010 - Sustainability and Corporate Responsibility Report. Retrieved July 23, 2012, from www.monsanto.com/SiteCollectionDocuments/2010-csr-report.pdf

11 Bollgard is a technology developed by the Monsanto Company to provide 'in-seed' protection to the cotton plants from those insects that attack the cotton bolls.

12 *Faces of Farming*, Monsanto Company,2010 Annual Report. Retrieved July 23, 2012, from www.monsanto.com/investors/Documents/Pubs/2010/annual_report.pdf

13 Ibid.

14 See an article on the website of Gene Campaign by Suman Sahai on *The Economics of Bt Cotton*, Gene Campaign Organization. Retrieved July 26, 2012, from www.genecampaign.org/Publication/Article/BT%20Cotton/BtCOTTON-Economics.pdf

15 Priya Kumar. (2009, June 2). "Biopiracy, GM Seeds and Rural India", *Centre for Research on Globalization*, p. 6. Retrieved May 2, 2012, from http://globalresearch.ca/index.php?context=va&aid=138

16 See the news report by Snehlata Shrivastav. (2011, August 28). "Reaping Gold Through Bt Cotton", *The Times of India*. Retrieved July 28, 2012, from http://articles.timesofindia.indiatimes.com/2011-08-28/special-report/29937803_1_bt-cotton-cry1ac-bollgard-ii

17 See the news report by Snehlata Shrivastav (2011, August 28). "Reaping Gold Through Bt Cotton", *The Times of India*. Retrieved July 28, 2012, from www. thehindu.com/multimedia/archive/01078/A_facsimile_of__i__1078418a.pdf

18 See the news report by P. Sainath on "Reaping Gold Through Cotton and Newsprint", *The Hindu*, Opinion, Op-Ed. Retrieved July 28, 2012, from www. thehindu.com/opinion/op-ed/article3401466.ece

19 See the news report by P. Sainath on "Reaping Gold Through Cotton and Newsprint", *The Hindu*, Opinion, Op-Ed, Retrieved July 28, 2012, from www. thehindu.com/opinion/op-ed/article3401466.ece

20 See the report by Dinesh C. Sharma dated 6 March 2010 on "Bt Cotton Has Failed Admits Monsanto", *India Today*, New Delhi, Retrieved July 5, 2012, from http:// indiatoday.intoday.in/story/Bt+cotton+has+failed+admits+Monsanto/1/86939.html

21 See the report by Dinesh C. Sharma dated 6 March 2010 on "Bt Cotton Has Failed Admits Monsanto", *India Today*, New Delhi, Retrieved July 5, 2012, from http:// indiatoday.intoday.in/story/Bt+cotton+has+failed+admits+Monsanto/1/86939. html

22 See an article by Vandana Shiva dated 5 July 2012 "From Seeds of Suicide to Seeds of Hope: Why Are Indian Farmers Committing Suicide and How Can We Stop This Tragedy?", *Huff Post World, The Internet Newspaper: News Blogs Video Community*. Retrieved July 6, 2012, from www.huffingtonpost.com/vandana-shiva/from-seeds-of-suicide-to_b_192419.html

23 See an article by Vandana Shiva dated 5 July 2012. "From Seeds of Suicide to Seeds of Hope: Why Are Indian Farmers Committing Suicide and How Can We Stop This Tragedy?", *Huff Post World, The Internet Newspaper: News Blogs Video Community*. Retrieved July 6, 2012, from www.huffingtonpost.com/ vandana-shiva/from-seeds-of-suicide-to_b_192419.html

24 See an article by Vandana Shiva dated 5 July 2012. "From Seeds of Suicide to Seeds of Hope: Why Are Indian Farmers Committing Suicide and How Can We Stop This Tragedy?", *Huff Post World, The Internet Newspaper: News Blogs Video Community*. Retrieved July 6, 2012, from www.huffingtonpost.com/ vandana-shiva/from-seeds-of-suicide-to_b_192419.html

25 See an article by Vandana Shiva dated 5 July 2012. "From Seeds of Suicide to Seeds of Hope: Why Are Indian Farmers Committing Suicide and How We Can Stop This Tragedy?", *Huff Post World, The Internet Newspaper: News Blogs Video Community*, Retrieved July 6, 2012, from www.huffingtonpost.com/vandana-shiva/from-seeds-of-suicide-to_b_192419.html

26 Ibid.

27 Monarch butterflies are large, beautifully coloured butterflies. They can easily be recognised by their striking orange, black and white markings. The most amazing thing about them is that they undertake an enormous migration each year.

28 The ladybird is a kind of insect also called Coccinellidae. It is a family of small beetles of size 0.8 to 18 mm. They are commonly yellow, orange or red with small black spots on their wing covers, with black legs, heads and antennae. They are considered agriculturally useful insects, as they prey on harmful agricultural pests.

29 See an article by P. Chengal Reddy dated 27 November 2001. "Fictional Agriculture of Environmentalists", *The Hindu*, November, Retrieved July 6, 2012, from www.hindu.com/thehindu/op/2001/11/27/stories/2001112700290100.htm

30 See an article by P. Chengal Reddy dated 27 November 2001. "Fictional Agriculture of Environmentalists", *The Hindu*, November, Retrieved July 6, 2012, from www.hindu.com/thehindu/op/2001/11/27/stories/2001112700290100.htm

31 See an article by Jyotika Sood dated 7 June 2012 on "Bt Cotton Has Improved Farmers' Lives", *Down to Earth Magazine*. Retrieved July 21, 2012, from www. downtoearth.org.in/content/bt-cotton-has-improved-farmers-lives

32 See an article by Jyotika Sood dated 7 June 2012 on "Bt Cotton Has Improved Farmers' Lives", *Down to Earth Magazine*, Retrieved July 21, 2012, from www.downtoearth.org.in/content/bt-cotton-has-improved-farmers-lives

33 See the news report by Zia Haq dated 26 March 2012. "Ministry Blames Bt Cotton for Farmer Suicides", *Hindustan Times*, Retrieved July 6, 2012, from www.hindustantimes.com/News-Feed/Business/Ministry-blames-Bt-cotton-for-farmer-suicides/Article1-830798.aspx

34 Ibid.

35 See the information provided by Sudhir Kumar Kaura on 18 June 2010 about "Bt Cotton Killing Animals in North India on Large Scale", *Indymedia Scotland*, Retrieved July 6, 2012, from www.indymediascotland.org/node/19734

36 "The Thirty-seventh Report of the Committee on Agriculture (2011–2012) on 'Cultivation of Genetically Modified Food Crops – Prospects and Effects' pertaining to the Ministry of Agriculture (Department of Agriculture and Cooperation) was presented to Lok Sabha and laid on the Table of Rajya Sabha on 09 August, 2012" (Committee on Agriculture-59th Report, 2013–2014, pp. 7). The Action Taken Replies by the Government on the Report were received on 30 November 2012, based on which final report was prepared.

37 See the biography of Dr Pushpa M. Bhargava on http://pmbhargava.com/pushpa-mittra-bhargava/

38 A type of plant that has been deliberately developed to have particular features.

4 Theoretical analysis of the policy development around GM crops

Introduction

From among the various theories on policy processes, this book adopts the advocacy coalition framework (ACF) theory in an attempt to analyse the issues and questions related to GM technology raised in the preceding chapters. The ACF theory was developed by Sabatier and Jenkins-Smith. It broadly explains the way in which political conflict takes shape after the emergence of a new issue and how players or stakeholders look to influence policy matters in this regard. In other words, the aim of ACF is to understand the political hurdles that come in the way of adequate utilisation of GM technology, which has become a burning issue, and the manner in which the interests of state, civil society and the corporate sectors are affected.

Rationale for adopting ACF theory

There are several theories that can help in developing an understanding of the GM policy debate in India. Let us look at a few such theories.

Theory of pluralism

According to the classical theory of pluralism, decision-making is usually the responsibility of the government, but the government can be influenced by many non-governmental groups. The central question of pluralism is 'how power and influence are distributed in a political process'. As attaining power is a continuous bargaining process between competing groups, groups of individuals try to maximise their interests, and lines of conflict that are numerous are always shifting. This is the result of conflict, bargaining and coalition formation among a potentially large number of societal groups motivated to protect and advance common interests. According to Mark Turner and David Hulme (1997), pluralism is an idealised model of western democracy holding the belief that "power is widely distributed among a variety of groups and channels for grievances are numerous and open". This notion is especially held widely in the United States. The state largely acts as an arbiter in this democratic competition while responding to the pressures coming from the society (Turner & Hulme, 1997).

Public choice theory

This theory deals with some traditional problems of political science, and it includes the study of political behaviour. This theory holds that politicians, bureaucrats and voters are self-interested, opportunistic and maximisers. It has kinship with the pluralist approach in its basic assumption that political society is composed of organised interests (Turner & Hulme, 1997). Critiques of this theory hold that public choice sometimes entails the furtherance of narrow interests instead of public interests. Here the weak and poor have the chance to become losers until they can organise themselves to articulate their interests. The only difference between pluralism and public choice is that while the former sees wise policies resulting from competing interest groups, the latter has no illusion about the cynical and self-interested character of actors (Turner & Hulme, 1997).

Rational choice theory

The pure form of rational actor model involves a sequence in which goals are identified, translated into objectives and then finally prioritised. This theory again has a kinship with the theory of pluralism in the sense that actors (whether persons, governments or other agencies) behave as rational choosers between alternative sources of action. The rational approach does not assume the actor's preoccupation with self-interest that takes the central role in public choice theory. Its approach is considered to belong to a perfect and ideal world where there are no constraints in time, resources and knowledge. It is simply an ideal model that can never be achieved in a changing and dynamic real world (Turner & Hulme, 1997).

The network approach

In this approach, political processes are not controlled by state actors alone; rather, they are characterised by the interactions of public and private actors. The concept of a policy network has several roots. Initially, it was strongly influenced by the "inter-organisational theory, which stresses that actors are dependent on each other because they need each other's resources to achieve their goals" (Adam & Kriesi, 2007). This is the central idea at the core of most approaches to the network. Later on in political science, the concept of a policy network evolved. The policy network represents an intuitively comprehensible metaphor, like regular communication and frequent exchange of information, that leads to the establishment of stable relationships between different actors and the coordination of their mutual interests. Therefore, it constitutes a new form of governance characterised by the predominance of informal, decentralised and horizontal relations (Adam & Kriesi, 2007).

Although each of these theories has helped in the development of one perspective or another, each of these theories cannot alone describe the complications involved in the policy debate on GM crops. For the appropriate description and understanding of the complications and dynamics of GM technology, an amalgamation of all the theories mentioned is required.

Therefore, the ACF developed by Sabatier and Jenkins-Smith seems suitable, as it is able to draw an explicit picture of the political conflict influencing policy processes among several important players. The aim is to understand through ACF the political hurdles that come in the way of adequate utilisation of the new technology and the manner in which the interests of state, civil society and corporate sectors coincide and clash. With the help of ACF, efforts are being made to understand the formation and deformation of associations with the emergence of new technology issues. To begin with, ACF will be analysed at greater length here.

The advocacy coalition framework theory

The ACF theory can be understood at three levels, i.e. 'macro', 'micro' and 'meso' levels, which are considered to be the three foundation stones. A flow diagram of ACF theory (Figure 4.1) can help in understanding these three essential levels.

Policy subsystem and external factors (macro level)

The theory assumes that policymaking in modern society is a highly complex process and therefore representatives of any stakeholders or interest groups must specialise in the area they intend to influence. According to Sabatier and Weible (2007), any subsystem is characterised by both territorial (e.g., civil society) and functional (e.g., policy making bodies) dimensions. The policymaking participants include not only the 'traditional iron triangle' of legislators, bureaucrats and judiciary but also researchers and journalists who specialise in the policy area. The theory further assumes that policy participants hold strong beliefs and are motivated to translate those beliefs into actual policy. Their beliefs are assumed to be strong and stable, and this makes a major policy change within the system difficult (Sabatier & Weible, 2007).

According to Sabatier and Weible (2007), in developing countries many subsystems are quite nascent because of the instability of the broader political system and lack of trained personnel in the subsystem. They argue that the majority of policymaking occurs within policy subsystems where specialists

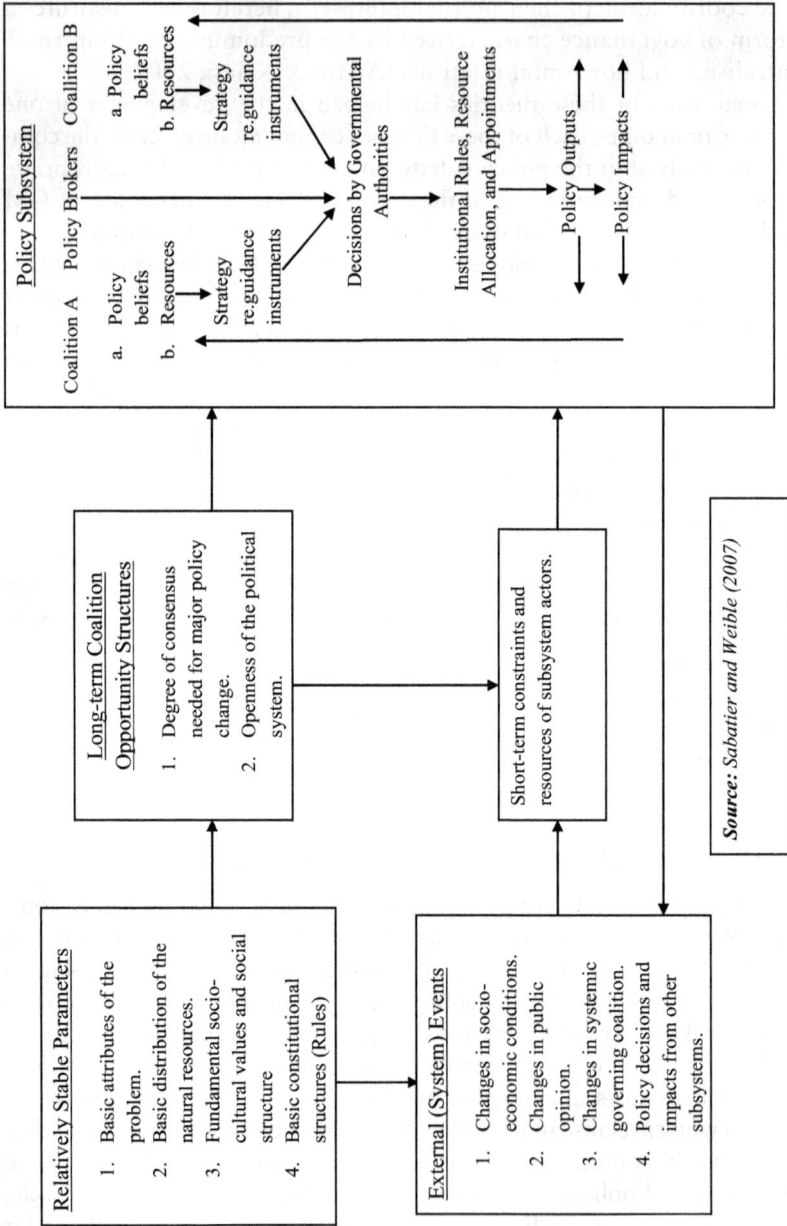

Figure 4.1 The ACF flow diagram drawn by Sabatier and Jenkins-Smith

Source: *Sabatier and Weible (2007)*

are involved in negotiations. However, the behaviour of most of the policy participants in the subsystem is mainly affected by two sets of exogenous factors, one of which is fairly stable and the other which is quite dynamic (see Figure 4.1). The relatively stable parameters include the basic attribute of the problem, distribution of natural resources, basic constitutional structure and fundamental socio-cultural values and structure. These parameters rarely change even in a decade, and thus rarely provide any impetus for policy change within a policy subsystem. On the other hand, the dynamic external factors include changes in socio-economic conditions, changes in the governing coalitions and policy decisions from other subsystems. The ACF hypothesises that a change in at least one of these dynamic factors is a necessary condition for a major policy change.

The individual model and belief systems (micro level)

At this level, the developers of ACF theory intend to show how it is different from rational choice theory. In general, rational choice theory assumes self-interested actors rationally pursue relatively simple material interests. However, ACF theory assumes that normative beliefs must be empirically ascertained and therefore does not exclude the possibility of a priori altruistic behaviour. ACF stresses the difficulty of changing the normative beliefs of policy actors. According to this theory, each policy actor or participant sees the world through a set of perceptual filters composed of pre-existing beliefs that are difficult to alter. Therefore, it can be argued that actors from different coalitions are likely to perceive the same information in very different ways, leading to distrust.

The ACF also borrows a key proposition from the "prospect theory" developed by Quattrone and Tversky in 1988 (Sabatier & Weible, 2007). According to them, actors value losses more than gains. Drawing from the implications of this, the ACF argues that individuals remember defeats more than victories. These propositions further interact to produce the "devil shift", which is the tendency of actors to view their opponents as less trustworthy and more evil and powerful than they probably are. As a result, the density of ties among members within the same coalitions increases and exacerbates conflict across competing coalitions. ACF theory also maintains that perceptual filters of individual actors or interest groups tend to throw away all dissonant information and accept only conforming information, which further strengthens their beliefs and makes changing beliefs more difficult. Therefore, this individual model of ACF is well suited to explain the escalation and continuation of policy conflict.

Advocacy coalitions

According to the ACF, the behaviour and beliefs of any stakeholder are shaped by the particular society or groups to which they belong. For example, if an individual is a member of a farming community or civil society group like Navdanya and Gene Campaign, or medical scientists concerned with health issues or environmental groups like Greenpeace concerned with the environment, then their beliefs and behaviour will be shaped accordingly. When the stakeholders belonging to these groups and societies holding different beliefs become policy participants, they naturally tend to influence the policymaking processes. The ACF assumes that policy participants strive to translate the elements of their beliefs into actual policy before their opponents can do the same. Therefore, in order to have any prospect of success, they seek to form allies, share resources and develop complementary strategies. The "devil shift" exacerbates the fear of losing to opponents, thus motivating the actors to align and cooperate with allies. The ACF has argued that policy participants would generally seek allies with people who hold similar beliefs among legislators, agency officials, interest group leaders, judges, researchers and intellectuals from multiple levels of government. Further, to form an advocacy coalition, some degree of non-trivial coordination is required among the allies of policy participants. The ACF further shows that an advocacy coalition provides the most useful tool to aggregate the behaviour of hundreds of organisations and individuals involved in a policy subsystem over a period of a decade or more.

The political controversy around GM technology

The issue of GM technology became more contentious in India after Bt cotton was approved for commercial cultivation in 2002. This move by the Government of India (GOI) led to different political stands either for or against GM technology. With the help of ACF, the political controversy behind the GM crop technology can be analysed. Based on Figure 4.1, a contextualised Figure 4.2 is given.

The crux of Sabatier and Jenkins-Smith's ACF theory is that the formation of coalitions takes place among different interest groups either in favour of or against a particular issue in order to change policy formation in the future. In the previous case, the flow diagram represents the changing situations due to the introduction of new GM technology. Before the introduction of this technology, there was an existing system in accordance with conventional agricultural practices. Suddenly when GM technology was introduced, several concerns related to health, environment and ownership issues came to the forefront apart from the benefits from this technology. These issues gave rise to a new political discourse between the proponents and opponents of GM technology, and the new political identities were recognised in terms of associations with the formation of new coalitions. The coalitions are formed on the basis of similar beliefs about the new GM technology among different stakeholders. The stakeholders can be biotechnology scientists, medical scientists, farmers, environmentalists, business groups, etc. The priorities of the different

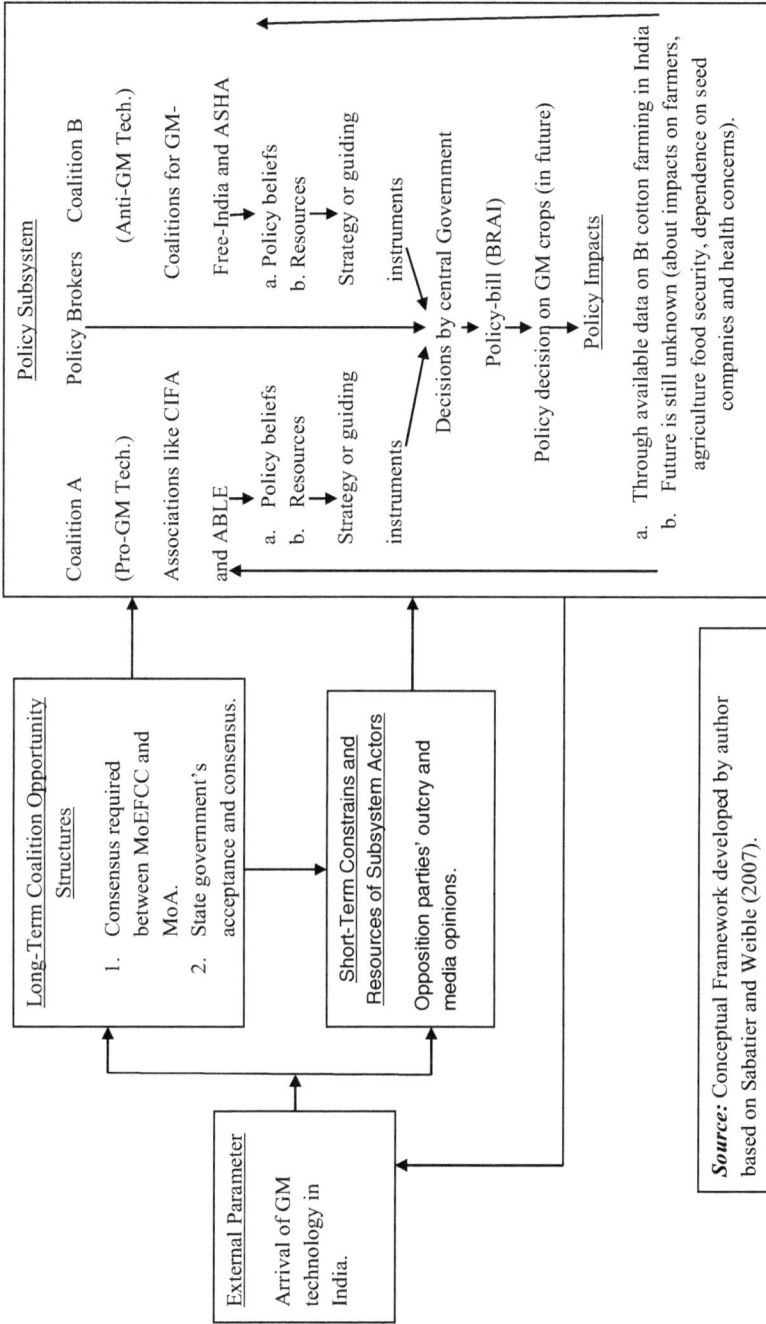

Figure 4.2 The flow diagram of the policy process of GM technology based on ACF

stakeholders are different. For example, the priority of a medical scientist may be providing health, which can be different from the priority of a plant biotechnology scientist who wants to enable plants or crops to grow in adverse weather conditions by manipulating their genes in labs. Similarly, the priority of environmentalists would be to preserve biodiversity. So based on different priorities, the demands raised by these stakeholders would differ and thus give rise to a discourse on how to deal with GM technology and how it can accommodate their different priorities. The different voices and concerns need not always result in a stand-off or adversarial position.

Keeping this argument in mind, four coalitions have been identified in the Indian context, which are either in support of or against the use of GM technology in agricultural production. These coalitions can be categorised as shown in Table 4.1.

In Figure 4.2, two coalitions, viz. coalition A and coalition B, are shown. Coalition A is in favour of GM technology, and coalition B is anti-GM technology. The civil society groups who belong to coalition A are Shetkari Sanghatana, Punjab Agricultural University (PAU) Kisan Club, Naujawan Kisan Club, Nagarjuna Rythu Samakhya and Pratapa Rudra Farmers Mutually Aided Coop Credit and Marketing Federation.[1] These organisations are among the leading farmer-supported organisations that are led by the Consortium of Indian Farmers Association (CIFA)[2] protesting the recommendations of the Supreme Court–appointed Technical Expert Committee (TEC) for a moratorium on field trials of GM crops, arguing that the farmer communities need biotechnology to increase agricultural production.

Apart from these civil society groups, there are also some corporate-sector groups representing the interests of Mahyco-Monsanto, Advanta, Kaveri Seed, Dhanuka Agritech, etc., that favour the government's decision to allow field trials of some 200 transgenic varieties of GM seeds. Following this decision of the government, biotechnology companies are getting ready with big investment plans for research and development (R&D). In order to promote GM technology, biotech industries have formed the Association

Table 4.1 Possible coalitions on GM technologies in India

Coalitions formed to support GM technology	*Coalitions formed to oppose GM technology*
a. Consortium of Indian Farmers Association (CIFA)	a. Coalition for a GM-Free-India
b. Association of Biotech-led Enterprises-Agriculture Group (ABLE-AG)	b. Alliance for Sustainable and Holistic Agriculture (ASHA)

Source: Prepared by the author

of Biotech-led Enterprises-Agriculture Group (ABLE-AG), in which biotech companies like Monsanto and Advanta are active members.

The scientists have mostly been in favour of using GM technology for Indian agriculture. Most of the plant biotechnology scientists of government institutes, such as the Centre for Cellular and Molecular Biology (CCMB) in Hyderabad, National Bureau of Plant Genetic Resources (NBPGR) of Indian Council of Agricultural Research (ICAR) in New Delhi, International Centre for Genetic Engineering and Biotechnology (ICGEB) in New Delhi and National Institute of Plant Genome Research (NIPGR) in New Delhi, believe that GM technology has the potential to increase agriculture production, especially from conventional crops yielding less and from those on the verge of extinction. The technology has the potential to enable crops to grow even in adverse weather conditions. For example, some GM maize crops have been designed in the lab to increase their tolerance level so that they can survive longer in drought regions. Therefore, according to the scientists of these institutes, GM technology has immense potential, but so far, this has not been realised because of several hurdles such as the protests against field trials of GM crops by anti-GM activists.[3]

Some of these anti-GM activists include civil society groups like Gene Campaign, Navdanya, Greenpeace, Andhra Pradesh Vyavasaya Vruthidarula Union (APVVU), Shaswat Sheti Kriti Parishad (SSKP),[4] Thana (an environmental organisation in Kerala), CREATE and FEDCOT (consumer rights groups), etc., and they are all protesting the government's move to promote GM technology in India. Some civil societies like non-governmental organisations (NGOs) and farmers' organisations have come together to form an association called 'Coalition for GM-Free-India'. This is a loose and informal network of scores of organisations and individuals from across India campaigning for a GM-free India and seeking to shift farming towards a sustainable path. Consisting of farmers, consumers, environmentalists, women and other organisations, this network is opposed to the environmental release of genetically modified organisms (GMOs).

Similarly, the Alliance for Sustainable and Holistic Agriculture (ASHA) is an alliance[5] of hundreds of organisations and individuals, including numerous farmers' groups, from more than 20 states of India and works on promoting sustainable agriculture and sustainable farm livelihoods.

Both these coalitions, which are either for or against GM crop technology, are trying to influence the Indian government in the policymaking process. The tension over the issue has been observed mainly between the Ministry of Environment, Forest and Climate Change (MoEFCC) and Ministry of Agriculture (MoA). The Genetic Engineering Appraisal Committee (GEAC), a statutory body under the ambit of MoEFCC, is entrusted with the responsibility of making the final call on whether GM crops have to be grown on Indian farms or not. It has been observed that the approach of the MoEF under different ministers has been changing. When Jairam Ramesh and Jayanthi Natarajan were in office, they were apprehensive of giving approval to field trials to GM food crops and a moratorium on GM crops (Bt brinjal) was imposed 2010. Later, Veerappa Moily under UPA and Prakash Javdekar under NDA removed restrictions by approving GM crops for field trials. Since 2015, there has been no tension between MoEFCC, MoA and Ministry of Science and Technology (MoS&T), as all these ministries have arrived at a common decision to approve GM crops for field trials to see whether they can help increase agricultural production in India (Navneet, 2014, 2018).

In India, agriculture is a subject that comes under the state list under Schedule 7, Article 246 of the Constitution. Therefore, the central government cannot alone decide on the matter of GM crops, and the state governments too have an active role in this regard. In Figure 4.2 the flow diagram, which is based on ACF theory, explicitly shows that if the system has to be transparent and open, then the opinions of state governments as well as the media and the opposition would also have to be considered.

Limitations of the ACF theory

The ACF flow diagram strives to explain the formation of new political identities and stands that either support or oppose GM technology. However, the theory is lagging in explaining the basic reason for the formation of such positions. Nevertheless, it does explain Maarten Hajer's argument that policymaking often takes place in a context where fixed political identities and stable communities are assumed and that policy interventions can make people aware of what they feel attached to by influencing their sense of collective identity, i.e. the awareness of what unites them and what separates them (Hajer, 2003). However, it does not throw much light on the question why new stands and identities are formed and the priorities of individuals or groups of individuals who identify with these new identities (Navneet, 2014, 2018).

Therefore, to adjust for this limitation of the ACF theory, the "co-dynamic model" has been adopted to understand the different priorities of various interest groups involved in either supporting or opposing GM technology. The very purpose of using the co-dynamic model is to study in detail one particular aspect of ACF theory, which shows two kinds of coalitions, Coalition A and Coalition B, taking shape in the case of GM technology. While Coalition A supports it, B opposes it, but both seek to influence the policymaking bodies of the government through various regulatory channels such as the Review Committee on Genetic Manipulation (RCGM) and GEAC.[6]

The co-dynamic model was developed by Millstone (Millstone, 2014). In order to understand the importance and reliability of this model, it would be necessary to understand 'the technocratic model' and 'the red book model', which represent the positivistic and scientific approaches towards policymaking processes (Millstone, 2014). Through these models, Millstone has attempted to analyse the reason for the conflict over GM technology in a given society. According to him, if a scientist favours GM technology, then he or she is looking for the solution of a specific problem related to agricultural yield, which would be different from those who are looking at the problem from the point of view of biodiversity or environment or health or ownership-related issues. With the help of Millstone's argument, it would be interesting to analyse and contemplate the priorities of the four coalitions – CIFA, ABLE-AG, Coalition for GM-Free India and ASHA – that intend to either support or oppose GM technology.

Therefore, the three specific models that are analysed here are:

1   The technocratic model
2   The red book model
3   The co-dynamic model

The technocratic model

The scientific expertise in policymaking emerged in the work of two positivists named Saint Simon and Comte in the nineteenth century in France. They stated that scientific knowledge is sufficient for policymaking, and their work later came to be known as the 'technocratic model' (Millstone, 2014). This model has appealed to expert scientific advisors because it has helped them in giving a narrative to depoliticise controversial policy issues and has further provided credibility to their knowledge, expertise and influence.

According to this model shown in Figure 4.3, scientific facts are self-sufficient in determining policy decision-making. The advocates of this model believe that a scientific approach is most appropriate for bringing about progress, as it can help shape more accurate and adequate policies.

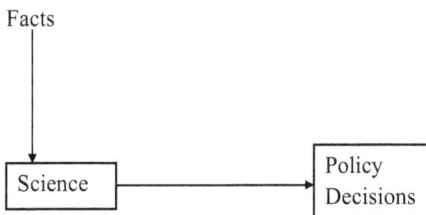

Figure 4.3 The technocratic model

Source: Millstone (2014)

They consider that policymaking should be based on 'sound science'. The implication of this model is that the responsibility of policy decisions should rely on technical and scientific experts, and the different ministries of the government should confine themselves to taking advice from these experts.

The red book model

The red book model is simply an improvised form of the technocratic model. It is a two-stage process model in which the first part deals with 'risk assessment' and the second part with 'risk management'.

Of the two stages, i.e. risk assessment and risk management, the first one is deemed to be a purely scientific stage and the second one a policy-making stage which has to deal with non-scientific and often normative considerations such as economic, social and political factors. In this two-stage model, the policymakers, who are also considered to be the 'risk managers', are well informed and influenced by the scientific advisors. The important fact to note here is that the scientific advisory bodies are portrayed in this case as entirely free from all kinds of influences from non-scientific elements. As shown in Figure 4.4, this model is linear and unidirectional. This means that the scientific advisory body would be in a position to influence the policymaking governing bodies, and the scientific advisory body would be free from any influences and biases and therefore their decisions would be taken seriously for shaping policies by the governing bodies. The model was first outlined in a report, 'Risk Assessment in the Federal Government: Managing the Process', drafted by the US National Research Council (NRC) in 1983. Because of the colour of its cover, the report came to be known as the red book (Millstone, 2014).

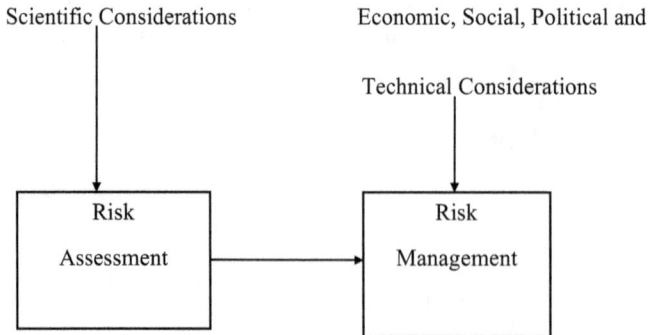

Figure 4.4 The red book model

Source: Millstone (2014)

Critique of the technocratic and red book models

According to Millstone, despite the official popularity of these models, they have been severely criticised by science policy analysts and sociologists with scientific knowledge. Both the technocratic and red book models presuppose that the available scientific knowledge is reliable and certain and that all experts can easily reach consensus, an assumption that is immensely false and baseless. In reality, the available science is often incomplete, ambiguous and uncertain, and therefore scientific communities have rarely been observed speaking with one voice.

Second, these models also presuppose that scientific assessments of risks and benefits can be achieved in highly neutral settings free from politics and social and cultural ethics. However, numerous scholars have already demonstrated several ways in which political, social, economic and cultural considerations have influenced the agendas, deliberations and conclusions of certain official scientific advice.

A co-dynamic, linear, bidirectional model

In order to solve the difficulties found in these models, a co-dynamic, linear, bidirectional model was derived as shown in Figure 4.5. Like the previous two models, this one is also linear, but unlike them, its starting point is not scientific facts and assessments, but a set of normative judgements about what is more important and which policy aim and objective should be pursued, priority wise. Second, unlike the previous two models, it is

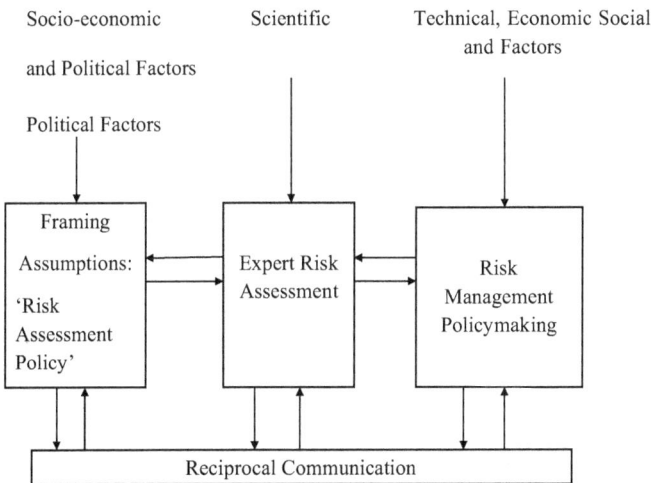

Figure 4.5 A co-dynamic model
Source: Millstone (2014)

not unidirectional but bidirectional and it is characterised by reciprocal interactions.

According to Millstone, this model assumes that science-based technology policymaking relies on both expert scientific assessments and non-scientific considerations. Instead of portraying risk assessment occurring in a policy-free space, the model represents scientific deliberations to be sandwiched between two sets of judgements. The first set of judgements guides what is to be assessed and thus seeks scientific answers to particular questions. In the second set of judgements, appropriate actions are thought of in the light of scientific answers, including comparisons with other alternative courses of action along with acceptability of affordable costs and benefits.

With the help of the co-dynamic model, Millstone explains how different groups of scientific risk assessors reach different conclusions. According to him, the reason for reaching a different conclusion is because the assessors asked and answered altogether different questions and on that basis reviewed different sets of data. This has further given rise to political conflict among different groups and stakeholders.

The co-dynamic model, along with the other two models, i.e. technocratic and red book model, can be taken and operationalised to understand the political controversy surrounding GM crops in India. The description of the other two models on which the co-dynamic model is built on simply enhances the importance of the co-dynamic model, making it more feasible for a democratic society where the voices of different stakeholders are taken into consideration in the policymaking process. Therefore, with the help of the co-dynamic model, the controversy around GM technology in Indian society can be analysed to understand the priorities of different stakeholders in either supporting or opposing it.

Both ACF theory and the co-dynamic model are complementary to each other. The theory explains the conditions and natures of different coalition groups formed on the basis of common beliefs and the manner in which different coalition groups engage in political conflict with each other. Similarly, Millstone's co-dynamic model helps in understanding the reason for the adoption of different priorities by different coalition groups.

The ACF theory and Millstone's co-dynamic model are useful in understanding the complexities of the GM crop issue in India. ACF theory talks about the formation of new groups in the form of coalitions after a new issue emerges. GM technology in this case has become a divisive issue among stakeholders because it is a new technology affecting Indian society. This chapter recognises the formation of four coalitions with the arrival of GM technology in India. These four coalitions are CIFA and ABLE in support of GM technology and the Coalition for GM-Free-India and ASHA against it. Within these four different coalitions, numerous groups of stakeholders are involved.

The ACF theory explains the basis for the formation of coalitions and further incorporates the propositions of other theories like the "prospect theory" developed by Quattrone and Tversky in 1988, which talks about actors valuing losses more than gain. These propositions further interact to produce the "devil shift", which is the tendency of actors to view their opponents as less trustworthy and more evil and powerful than they probably are. This increases the density of ties among members within the same coalition and exacerbates conflict across the competing coalitions. Therefore, with this kind of argument, ACF theory does provide the basis for understanding an existing political conflict among coalitions supporting or opposing a particular cause but it does not provide any reason why they do so.

From the ACF framework, one can very well understand how political conflict builds up among the given four coalitions. But one also needs to understand the basic reason why the stakeholders involved in these coalitions support and oppose GM technology. As already explained in the chapter, the stakeholders of GM technology are not only scientists, farmers and corporate companies like Monsanto but also consumer groups like CREATE and FEDCOT and civil society groups like Navdanya and Gene Campaign. For the development of effective policies in a democratic society like India, the voices of all the stakeholders are important and need to be considered. Millstone's co-dynamic model clearly explains the reason for the existence of various conflicting voices on the issue of GM technology. According to Millstone, the reason why the stakeholders arrived at different conclusions on GM crops is not because they are offering different interpretations of agreed and shared bodies of evidence, but because they have asked and answered different questions and therefore reviewed different data sets.

Keeping in mind the analysis made through the co-dynamic model, one can find that there is no agreed information available on GM technology. On the one side, official government data show that there has been an immense increase in cotton production after the cultivation of Bt cotton in Indian farms (Directorate of Economics and Statistics, 2012). At the same time, NGOs like ANTHARA have raised questions on whether GMOs are safe for animal and human consumption and whether sufficient testing has been done to know the safety concerns related to the health of animals and human beings. Other NGOs like Navdanya and Gene Campaign have raised environment and biodiversity concerns and have argued that GM technology is going to increase monoculture in the Indian agricultural system, which may further raise the risk of crop failures.

Therefore, the stakeholders have several priorities such as agricultural production, health, environment and ownership rights, and on that basis they are raising demands to influence policymaking, which is still evolving in the case of GM technology. So the development of effective social policy in a democratic society like India, agreeable to every party, is challenging and more time should be devoted to understanding the priorities of all stakeholders and accordingly utilise the benefits of GM technology for agriculture with a strict regulatory channel in place.

Methodology developed for the field survey

Based on ACF and the co-dynamic model, a field study has been under-taken to understand the priorities of different stakeholders who are part of the coalition groups discussed in Table 4.1 and the reason for their priorities. The field study has been planned in three phases. In the first phase, all those stakeholders directly involved in influencing the decisions of the policymakers in the ministry and regulatory bodies were interviewed. These stakeholders include biotechnology scientists; agricultural scientists; medical scientists; agricultural economists; some of the current as well as ex-members of the Indian regulatory bodies like GEAC; and members of civil society groups like Gene Campaign, Navdanya, Anthara, Centre for Sustainable Agriculture, Swadeshi Jaga-ran Manch, Bhartiya Kisan Sangh, etc. All of them are policy elites of their own respective organisations. To interview them an open-ended questionnaire was prepared so that the respondents can be engaged in informal discussions.

In the second and third phases of the field studies, a closed-ended ques-tionnaire was prepared to conduct a survey of 100 cotton farmers each in Telangana and Maharashtra to know their priorities and choices in terms of Bt cotton cultivation. Locations in Telangana and Maharashtra were selected for the survey, as both states have large areas under cotton cultivation and a large number of farmers have reportedly committed suicide in these states because of agrarian distress. So it was considered important to see how far GM technology in the form of Bt cotton helped the farmers in these regions in coming out of agrarian distress.

In the Telangana region, Warangal district was selected, as most of the farmers cultivate cotton in the region, which has a mixture of red and black soil conducive for the growth of cotton. Within the Warangal district, two blocks were selected for the survey: Hasanparthi and Atmakur. The total size of the sample is 100 farmers from both blocks taken together. From each block, 50 farmers were selected to answer a set of questions in a closed-ended questionnaire. Figure 4.6 shows the field survey design to conduct interviews with farmers in Telangana.

In Maharashtra, the Yavatmal district, which falls in the region of Vidarbha, was selected. It has reported the highest number of farmer suicides. Most of the farmers in this region are again cotton growers. Within Yavatmal district, two blocks were selected for the field survey: Ghatanji and Pandharkawda. The total size of the sample is 100 from both the blocks taken together. From each block, 50 cotton farmers were selected to answer a set of questions in a closed-ended questionnaire. Within each block, four villages were selected, and within each village, 12 to 13 farmers were chosen through a snowball sampling methodology for the survey. Figure 4.7 shows the field survey design to conduct interviews with farmers in Maharashtra.

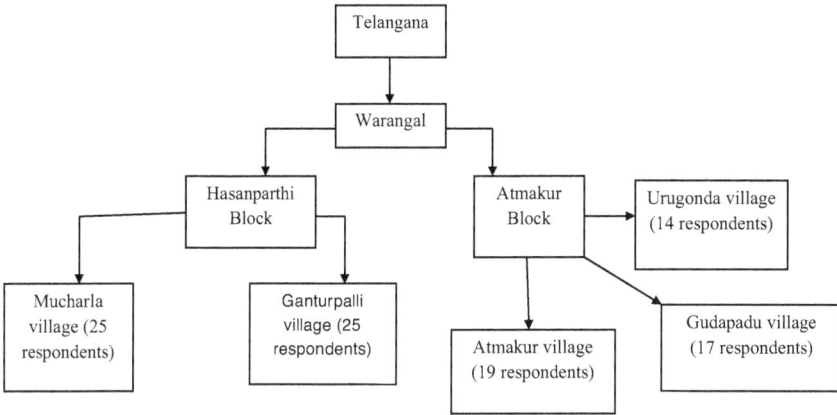

Figure 4.6 Field survey design for Telangana

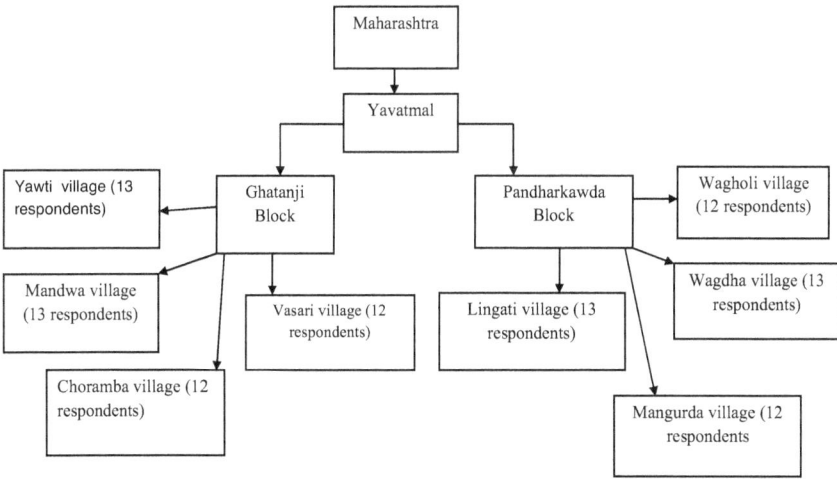

Figure 4.7 Field survey design for Maharashtra

The closed-ended questionnaire has been designed to look for answers from farmers in both Telangana and Maharashtra on the impact of Bt cotton cultivation on yields, returns, seed usage and expenditure, pesticide usage and expenditure, pest attacks, availability of irrigation facilities, reason for farmer suicides, human health, livestock health, soil quality and impact on the environment.

Notes

1 Shetkari Sanghatana is a non-political union of farmers formed with the aim of creating freedom of access to markets and technology for farmers. It was founded by Sharad Anantrao Joshi in the late 1970s.

  Similarly Punjab Agricultural University (PAU) Kisan Club was started by Dr T. S. Sohal in 1966 in Barewal village near Punjab Agricultural University. Like Shetkari Sanghatana, PAU Club is also a non-political, non-profit organisation of progressive farmers of the state. Similarly, Naujawan Kisan Club is a non-political and non-profit organisation based in Punjab, and Karnail Singh is its president. The Nagarjuna Rythu Samakhya and Pratapa Rudra Farmers Mutually Aided Cooperative Credit and Marketing Federation are the other two non-profit organisations of farmers based in Andhra Pradesh and Telangana regions, respectively.

2 CIFA was launched in March 2000 at Tirupathi by a leader of farmers, Shri Sharadh Joshi. Presently, it is functioning under the direction of Shri P. Chengal Reddy. CIFA is a coalition of different small organisations of farmers that work to enable Indian farmers to get access to modern technologies that can help them increase farm production and get direct access to the market.

3 Based on interviews conducted by the author with the scientists of the mentioned biotechnology institutes.

4 Gene Campaign is a research and advocacy organisation dedicated to the food and livelihood security of rural and Adivasi communities and the rights of farmers and local communities. It works with communities in villages as well as at policymaking levels to ensure the rights of farmers and local communities to biodiversity and indigenous knowledge. Gene Campaign was set up in 1933 by Dr Suman Sahai and a group of scientists, environmentalists and economists who were alarmed by the impact of international developments on the genetic resources of the developing world and the livelihood security of rural and tribal communities that depend on them.

  Navdanya started as a programme of the Research Foundation for Science, Technology and Ecology (RFSTE), a participatory research initiative founded by the world-renowned scientist and environmentalist Dr Vandana Shiva, to provide direction and support to environmental activism. The main aim of the Navdanya biodiversity conservation programme is to support local farmers and rescue and conserve crops and plants being pushed to extinction and make them available through direct marketing.

  Greenpeace India has been working on various issues related to the environment since 2001. It is a non-profit organisation with a presence in 40 countries across Europe, the Americas, Asia and the Pacific. Their work in India is focused on four broad campaigns, viz. stop climate change, ensure sustainable agriculture, preserve the oceans and prevent another nuclear catastrophe.

  Andhra Pradesh Vyavasaya Vruthidarula Union (APVVU) is a federation of unions of agricultural workers, marginal farmers, fisherfolk and rural workers in Andhra Pradesh. It was established in 1991.

  Shashwat Sheti Kriti Parishad *(SSKP)* is a farmers' organisation promoted by the Yuva Rural Association (YRA). The organisation has been taking up local issues such as availability of water and seeds. SSKP has also protested against genetically modified organisms (GMOs).

5 There is not much difference between the words alliance and coalition in terms of meaning; however, alliance represents a stronger group in terms of working together to achieve a specific goal.

6 Both RCGM and GEAC are the two important Indian biosafety regulatory bodies involved in the process of granting approval to GMOs. RCGM is an important body of the Department of Biotechnology (DBT), which falls under the ambit of the Ministry of Science and Technology. Similarly, GEAC falls under the Ministry of Environment and Forests (MoEF). RCGM mainly oversees research studies on GMOs, whereas GEAC is responsible for the approval of activities involving large-scale use of GMOs and their release into the environment through experimental field trials.

5   Analysis of the arguments of policy elites supporting or opposing GM crops

Field-based study in Delhi and Hyderabad

Introduction

This chapter deals with interviews done with policy elites. As mentioned in the previous chapter, the overall field study has been conducted in three phases. This chapter brings the details of the first phase of the field study. In the first phase, some of the renowned social and political elites whose organisations have been involved in influencing policy decisions on GM technology were interviewed. A total of 13 such elites[1] were interviewed. Among them are scientists and directors of reputed research institutes; current and ex-members of government-appointed GM regulatory bodies; an ex-national professor of the Indian Council of Agricultural Research (ICAR); and directors and activists working for civil society organisations like Anthara, Centre for Sustainable Agriculture (CSA), Navdanya, Gene Campaign, Bhartiya Kisan Sangh (BKS) and Swadeshi Jagaran Manch (SJM). Through the field study, an attempt has been made to understand the priorities and rationale behind decisions to support or oppose GM technology. For the study, an open-ended questionnaire was prepared and sufficient time for discussion was given to the interviewees to explain the reasons behind their stands.

Field discussions

It has been noticed that there exists agreement and disagreement among the stakeholders on whether Bt cotton seeds helped farmers in increasing cotton production. Monsanto's yearly reports and biotechnology scientists have been stating that with the introduction of Bt cotton seeds, there has been a tremendous increase in cotton production in the country. However, civil society groups have been arguing that cotton production has increased not because of the increase in yield per hectare but because of the increase in the net area under cotton production.

Dr Ramesh Sonti is a scientist at the Centre for Cellular and Molecular Biology (CCMB), and during the time of the interview, he was a member of one of the important organs of the Indian regulatory body, the Genetic

Engineering Appraisal Committee (GEAC). As a member of the regulatory body, he threw light on several concerns related to GM technology and its regulations. On whether Bt cotton seeds helped increase cotton production, Dr Sonti said:

> I feel that Bt seed has helped increase cotton production. But I will not say that all of the increase is only due to Bt. There are other factors also involved in helping in the increase in cotton production. Today the plant genotypes are much better than what it was 10 to 15 years back. These genotypes have improved because of the efforts put into the continuous breeding process. This can also be one of the factors in increasing cotton production. But in some locations Bt cotton has also helped in increasing cotton production. So instead of saying it is Bt cotton only that helped in the increase in cotton production or Bt cotton has not at all helped, I would say that the truth lies somewhere in the middle.[2]

On a similar note, Dr Pushpa M. Bhargava, who was the founder of the country's prestigious CCMB and had been appointed by the Supreme Court as a member of the GEAC, stated:

> The total amount of cotton production in India has increased. There is no question about this. But let us look at the reasons for this please. We have now a larger area under cotton. Therefore, the total production has increased. Bt gene has given resistance against certain pests. But Bt cotton seeds have not been successful in the rain-fed areas. In India, two-thirds of cotton cultivation is done in rain-fed areas. In rain-fed areas Bt cotton has not helped farmers one width. Lot of Bt cotton growers have committed suicide because of loss. These have happened mostly in the rain-fed and non-irrigated areas. Even where Bt cotton succeeded like in parts of Gujarat, slowly the insects have started becoming resistant against the Bt genes. So there the Bt cotton is not working anymore.[3]

Dr Virander S. Chauhan is a scientist at the International Centre for Genetic Engineering and Biotechnology (ICGEB), New Delhi, and has also served as a member of the GEAC. He argued that rain is a variable that cannot be figured out. According to him, the issue is not about genetically modified seeds, which are rain dependent, but rather about pest-resistant seeds. In his words:

> The pest resistance will be there whether it rains or doesn't rain. So that is immaterial. Now, one can raise another question, will the GM seeds take more water or not. If it takes more water than the normal crop, then we have a problem. But it does not take more water. Water

consumption doesn't go up or doesn't go down. So rain or any other variable like heat needs to be treated separately. The only variable we are pointing towards here is Bt enabling cotton plants to be pest resistant.[4]

Therefore, according to Dr Chauhan, even in rain-fed areas Bt cotton has performed better than non-Bt cotton. On the question of insects developing resistance against Bt cotton plants and Bt genes no longer being effective against insects, he argued:

Sure it will happen. It's like antibiotic treatment. Once there was a time when 50 to 60 per cent of penicillin used to work. Today penicillin has become useless. So you discovered a new kind of antibiotic. Second generation antibiotics don't work on us anymore. Today the major health problem is the antibiotic resistance. So this is nature. One can't do anything to the nature. If I produce everything with Bt, of course there will be bugs tomorrow. Now bugs have become resistant, so you do more research and find another. Bt is not the last roti (bread). No good scientist will say that tomorrow resistance will not take place. Let it come but research should always stay ahead in the whole game.

According to Dr Chauhan, resistance development in insects is not new. Scientists have been dealing with this, and they are prepared to deal with this through their research. To support his argument he further states:

Today in Japan and other places, the average human life span is 80. In India, it is going to be 67. This is because you discovered new drugs all the time. Today people live longer. Same thing will happen to the crops.

Unlike Dr Bhargava, Dr Chauhan feels that development of resistance is not problematic because scientists are prepared to develop the next generation of GM seeds if the present seeds become useless or fail to give protection to the plants against bugs.

Apart from agricultural production and resistance development, civil society groups have raised concerns regarding biosafety, biodiversity and health-related issues in GM crop cultivation. Vandana Shiva, who is the director of NGOs like Navdanya and the Research Foundation for Science, Technology and Natural Resource Policy, states that:

More chemicals don't produce more foods. They destroy the soil, they kill beneficial insects. Therefore, you have more pest attacks. They are destroying biodiversity through the use of herbicide round-up glyphosate.[5]

For Shiva, GM technology is a "new chemical" of the old paradigm of the Green Revolution. During the Green Revolution era, chemical industries earned huge profits by selling their pesticides for the new hybrid varieties of crops. GM technology is part of an old system with old interests. According to Shiva, it has not helped in replacing these chemical sprays. These industries have been selling their chemicals, and now with the help of GM technology, the corporations are in a position to own and control agricultural seeds as well.

India has signed several international protocols like the Cartagena Protocol on Biosafety (CPB) and has been a member of Codex Alimentarius Commission (CAC), which lays down guidelines for safety standards for food and aims at

> ensuring an adequate level of protection to the safe transfer, handling and use of living modified organisms (LMOs) resulting from modern biotechnology experiments and uses, which may have adverse effects on the conservation and sustainable use of biological diversity, and also taking into consideration risks to human health, and trans-boundary movements of LMOs across countries.
>
> (Final Report of the Technical Expert Committee, 2013, pp. 13–15)

Now, it was deemed important to ask scientists of both regulatory bodies and civil society groups how adequately the guidelines of these international protocols, to which India is a signatory, are being followed.

Therefore, on the question of whether biosafety protocols are being properly addressed by the regulatory bodies, Dr Sonti responded:

> There are sets of protocols followed by Department of Biotechnology (DBT) in consultation with Indian Council of Medical Research (ICMR) and ICAR for adequate safety assessments. Till now GEAC has not brought any change in the existing guidelines of the protocols but discussions for some change in the existing guidelines are in the process. The existing protocols have to be evaluated and then decided whether they need to be converted into the stringent one.

By this, Dr Sonti means that the biosafety protocols of the CPB and CAC remain mostly on paper and that discussions are going on among scientists regarding their implementation. On a similar note, Dr Chauhan, the ex-GEAC member, states that:

> I think in India as they exist, they are very strict. In fact, the bio-regulatory measures are very tight. Whether it can police or not is a different thing, but the legal framework or the framework of biosafety is very very strong, much stricter than many countries we know. But whether you can implement it or not, I am not sure. For example, we have very good

traffic laws in India, but if citizens don't follow the traffic rules and regulations then I can't do very much about it, except to catch you. So to go back to your question, I think the regulatory boundaries that have been given to us are very tight. Much more tighter than any countries.

Dr Chauhan argues that on paper, the biosafety framework is very strong but its implementation needs to be evaluated carefully. Dr K.C. Bansal, who is a scientist and director of the institute National Bureau of Plant Genetic Resources (NBPGR), during the time of interview stated that in this regard all biosafety measures are taken care of by the government regulatory bodies and all relevant national and international protocols have been put in place.

On a similar issue, Dr Ranjini Warrier, who at the time of the interview was the director of the GEAC, stated:

> Cartagena protocol is specific to the trans-boundary movement and it has three categories. One is for the research purpose, one is for the intentional release into the environment and the last one is for food, feed and processing only. So for each category, risk varies. The protocol guidelines don't advocate the same kind of precautions for all the three categories. If it is a research, then it is for the contained use. In that case it is not going to be released into the environment. Therefore, for that a different kind of document is required. But, if it is for the intentional release, you have to conduct a detailed risk assessment testing and other things. In our country, we don't allow import of GM crops for direct release into the environment. We only allow for contained use. As far as food, feed processing is concerned, we don't import any GM food because we have been till now self-sufficient in food. We have only allowed the import of GM soybean oil, which is a highly processed food.[6]

According to Dr Warrier, all the precautionary measures for food safety, as prescribed in the CPB, have been taken care of. Dr Bansal has stated in this regard that:

> The Government of India and her regulatory bodies are following all the international protocols to which India is a part and signatory member.[7]

However, the members of civil society groups have different opinions on this matter. They have alleged that the government regulatory bodies have not been adequately following the international protocols of CPB and CAC.

Dr Sagari R. Ramdas, who is a veterinary doctor and also a founding member of the NGO Anthara, has said regarding biosafety protocols that they are not in place. In her own words:

> We have Bt cotton everywhere and animals are grazing on it. So you should be having the local equivalence of Public Health Centre (PHC) in

the form of veterinary hospital. The veterinary hospital at the Panchayat or mandal or block level should be having all the systems in place, so that even if some animal dies, you are able to access the testing facilities available at the veterinary hospital in order to know the cause of the death. That's where it needs to be built in, otherwise nothing is in place. So, when you talk about biosafety, that means biosafety has to be there for any potential life forms with which genetically modified organism (GMO) will interact. But, you don't have mapping either in place. Among animals, birds, beasts, insects, humans, the interaction is very much there during consumption of food. But amidst of all this, we don't have anything in place to take care of biosafety mechanisms.[8]

During the field study, it was observed that several scientists and economists denied that there has been any health impact on animals feeding on cotton plants. They have argued that animals eating cotton plants might be falling sick because of chemical pesticides rather than the Bt gene in the cotton plants. To this, Dr Ramanjaneyulu, who is the director of Centre for Sustainable Agriculture, brings an interesting argument, pointing out that even if it is assumed that the health of the animals has been affected due to other reasons, nobody can deny the possibility of gene contamination in the fields because of GM crop cultivation. There is evidence to show that gene contamination has taken place even in the United States. He stated:

If you look at the US for example, two of the very popular and serious contamination cases are from the US actually. The wheat contamination case,[9] which happened in Oregon state and the Liberty Link rice contamination (LL rice) case,[10] which happened in Michigan. LL rice controversy and GM beef controversy both show that even a country like the US, which does not have diversity for crops, have faced serious problems. Therefore, India, which has huge biodiversity in almost all the crops, faces a serious threat from GM crops. That's the first thing I would say. Secondly, if you look at the health related issues, studies show that blood samples of new born babies had the traces of herbicide tolerant gene in DNA. So there are evidences available. The problem is that the mainstream system, be it the public or government institutions, all are turning a blind eye to the whole issue.[11]

With these arguments, Dr Ramdas and Dr Ramajaneyulu have shown that biodiversity and biosafety are closely linked. The job of the scientist, besides the development of technology to enhance agricultural production, is also to look into how different species, both among plants and animals, interact with each other. This is because all of them are dependent on each other forming an ecological balance. Therefore, disturbances in the population of one species can bring hindrance to the life cycle of other species as well.

Further, raising concerns over the health issues of animals grazing on cotton fields, Dr Sagari said that her organisation, Anthara's, engagement with GM crops began when shepherds, who used to rear sheep and goats, began to observe that the animals were suffering from unknown diseases after grazing on cotton crops. Some of them even died of unknown reasons. It was then that Anthara's team of veterinary doctors decided to do some fact finding in this regard. Dr Sagari says,

> Ever since we got this report of animals dying because of grazing on Bt cotton field, the first thing we did for the first three year was to go and do fact finding. In doing so, we would mostly reach out to the place where the animals were grazing on Bt cotton plants. We studied the case history of these animals for three consecutive years, i.e. 2005, 2006 and 2007. Mostly these animals used to be sheep. When animals used to show any symptom of disease, the shepherds used to take them to the local veterinary doctors. Post-mortem was done on the dead sheep to know the cause of their death. Samples were then sent to the veterinary and animal husbandry laboratories. We as an organisation also filed an RTI to know what exact samples were received by the laboratories and what response came from it thereafter and also what kind of tests were performed to come to the conclusion. However, to our sheer surprise, despite filing an RTI, we were unable to get any proper answers.

Both Dr Sagari and Dr Bhargava have alleged that scientists in the regulatory bodies have been analysing the health of the animals feeding on cotton plants through acute toxicity tests[12] rather than chronic toxicity tests.[13] Dr Bhargava has explained that:

> Scientists have done tests only for 90 days, which is not enough. They have checked for a few toxicities and that has also been done by the company itself. But chronic toxicity, which is a feeding study for long periods, have not been done in any experiment till now. There was a study done by Seralini a few years ago, showing that GM crop can lead to cancer and other problems. That was a long-term study done by feeding mice with GM food for the entire life time.

Dr Bhargava has said that very few studies have evaluated animals by feeding them GM foods for a long period. According to him, most of the studies have been done for a short period, looking only for acute toxicity. Dr Bhargava stresses here that:

> We need to know about chronic toxicity because after all, anything that causes cancer does not cause it in 10 or 90 days. It can take the life time of the animal to understand that.

According to Dr Bhargava, the animals should have been fed the suspect food for the entire two-year period, which is the average life span of rodents. Dr Bhargava has also alleged that

> In the country, we have no independent laboratory today. We only depend on the results produced by the company who themselves manufacture these technologies.

Therefore, according to Dr Bhargava, a clash of interests appears in this case. Companies producing the technology will not show anything negative about it in their data. Their interest lies in promoting it in order to make profit. Therefore, an independent assessment of these technologies needs to be done by the scientists. The members of civil society groups have been backing this view of Dr Bhargava and have been raising demands for the independent testing of GM technology by the regulatory bodies.

As a member of GEAC, Dr Sonti agreed that very little independent research has been done in the area and scientists at the regulatory bodies have been depending on the data provided mostly by the private companies, who manufacture and intend to promote these technologies. He said this is because testing requires huge funding, and since the government has not provided enough money to conduct independent testing, scientists at the regulatory bodies mostly rely on the data provided by the private companies. On a similar note, Dr Akhilesh Tyagi, a scientist who has served as a member of the GEAC and was the director of the National Institute of Plant Genome Research (NIPGR) during the time of interview, makes his point by saying:

> In scientific practice, we write research papers and scientists read them and they make their assessment about those papers. Now every individual who is reading that paper will not repeat that paper in order to say that this data is correct and this data is wrong. It comes with experience also. Some part of this is repeated if there is a requirement and is verified. Other part is seen as technical protocols that have been followed. So the problem is that if it is a matter of falsification, you would raise the question even if the independent person is doing the test. So where is it going to stop? Ultimately, it is expected that in all the regulatory bodies, there are scientists with experiences and for something, which can be monitored on the site, they do monitor.[14]

Dr Tyagi argues here that scientists who assess the data produced by the private companies are experienced and understand very well which data need to be checked by conducting experiments and which can be simply considered data. They do not need to reproduce the whole data once again to understand whether the data produced by a private company are correct or not. Scientists are well aware of the methodology to conduct experiments, and therefore, they just have to assess the various steps taken while test trials were conducted by the companies. By doing so, they do not have to repeat all the experiments already conducted by the companies in order to generate data.

Dr Chauhan argues that the government as a system can also be corrupt, and so it would be risky to depend completely on the government to control every matter. He says:

> It is a reality that members of the government regulatory bodies simply assess the data produced by the private companies regarding the field trials of the GM crops. If the companies are making any technologies, then it is their job as well to produce data regarding safety issues. Now first of all, you are either going to trust the system or not trust the system. If you don't trust the system, then you have to have your own system in place. But you might argue that the government system itself can be corrupt. Personally I think that government should not get involved in giving clearances. Our history will tell you that for seventy years, wherever you have put control, there has been corruption. It has happened in health and many other sectors. So what to do then? You have to trust the companies. Now if companies are lying, you will soon find that because the data is with you.

Dr Chauhan seems to say that if one cannot trust the private companies, one also needs to be cautious of the control by the government. He believes that even if the government controls everything, one cannot be sure that corruption will not take place. It can take place within the government system as well. Therefore, according to him, the best possible way is to appoint good, experienced scientists in the regulatory bodies who can understand and analyse the data generated by the companies regarding the safety trials of the technologies. He also suggested that these companies can be told not to collect data by themselves but to entrust the task to registered CROs,[15] which will further decentralise the overall process.

Therefore, the very idea of an independent scientist is very blurred. Every scientist is associated with some organisation or the other and therefore will have different views. The view of a scientist with an environmental science background will be different from that of a biotechnology or agriculture scientist. Dr Ramesh Chand, who was earlier an ICAR national professor and during the time of interview was the director of National Institute of Agricultural Economics and Policy Research (NIAP), says:

> Scientists' views are always divided on anything of any professional matter. Even on nuclear energy, views of scientists are divided. It is, I think, totally wrong on the part of anyone to presume that there will be consensus among scientists. I think then they cease to be scientists. In the professional world, there is no consensus. If there is a division among scientists, I am not surprised about it. It has always been there,

it is now and it will continue to be there. Many people say we should use chemical fertilisers, some people say we should not use them. In professional world, people are always divided and so it is not that only on GM crop they are divided.[16]

Therefore, the demand raised by the civil society that there should be an independent test by independent scientists free from any conflict of interest is questionable. It is difficult to find any scientist who is free from any kind of interest, and therefore, whatever scientific theory he or she will come up with would not be free from biases that cannot be questioned. On a similar note, Dr Suman Sahai, who is a scientist and the founding director of the NGO Gene Campaign, argues that:

> There is certainly conflict of interest among scientists. I think that a larger malaise in our scientific community, which is really tragic to observe, is that scientists have lost their independence. Scientists have become government employees. As a result of which, they are very reluctant to speak against the government policy. If the perception is that the government is promoting GM technology, which very clearly is, then scientists, even though in private may have concerns, they are very reluctant in public to voice it. Then there are other scientists, who are beneficiaries of grants support and who will not open their mouth because they are the beneficiaries of substantial means.[17]

Dr Sahai's organisation Gene Campaign has shown its dissent against the approval of field trials for some GM crops by the regulatory bodies. Representatives of civil society groups like Gene Campaign, Navdanya, CSA, SJM and BKS have been arguing that though proper conditions have not been established, regulatory bodies have approved the field trials. Dr Ashwani Mahajan, who is the spokesperson of SJM, says in this regard that:

> For conducting field trials, the kind of conditions which is required has not been fulfilled. They (scientists at the regulatory bodies) say that they will conduct field trials at an isolated place and construct adequate boundaries so that pollens won't be able to cross those boundaries, which is an immensely unscientific and illogical argument. The fact is that pollens can't be stopped and they can travel for hundreds of kilometres through air. So in India, the conditions are not ripe for conducting field trials.[18]

Dr Mahajan then raises the question "why can't those trials be done in glass houses rather than doing it in open air?" Dr Mahajan has alleged that the members of the regulatory bodies want to conduct field trials of GM crops even in the absence of adequate conditions and infrastructure, which is dangerous because the next step after field trials is commercialisation of the GM crop. According to him, they want to jump into the commercial application of this technology as soon as possible without examining closely biodiversity- and biosafety-related protocols. He has argued that many other options are available for increasing agricultural production. He cites the example of Israel and Bihar by saying:

> Look at Israel, how they managed to increase their agricultural pro-duction without using GM technology. In Bihar, rice production has increased with the help of better technologies rather than GM technol-ogy. So there are many ways to increase the production. There is a quote from M.S. Swaminathan that GM technology should be adopted only if other possibilities have been ruled out.

Therefore, Dr Mahajan believes that the country does not need GM technol-ogy to increase agricultural production. According to him, several alterna-tive possibilities are available to increase the food basket, so scientists should focus on making use of those possibilities rather than haphazardly creating a market for GM technologies.

On a similar note, Mr Mohini Mohan Mishra, who is the spokesperson of BKS, has criticised the approval of GM technology both for field trials and for commercial cultivation. He is severely critical of those scientists who compare GM technology with other technologies like cell phones, television, etc. He says these scientists hold the mistaken belief that in the same way cell phone technology has advanced in quick succession, so can GM crops and that if an older version of a GM crop fails against the attacks of bollworms, newer, more resistant models can be developed. But Mr Mishra argues:

> If something goes wrong with cell phones you can always take it back. This kind of technology is reversible, correctable and repairable. Whereas, GM technology is not reversible, correctable and repairable if it brings damage to the biodiversity of existing crops and environment. Right now, your Bollgard-I has stopped working. Therefore, Bollgard-II has been introduced. Now once Bollgard-II has been introduced, there is no possibility to bring back Bollgard-I. We cannot reverse the process. We cannot repair the damage that has been brought to the biodiversity till today and that will be brought by GM technology in future. So this is foolish to compare GM technology with other technologies like cell phones or televisions.[19]

Mr Mishra believes that GM is very different in comparison to other tech-nologies in the sense that once it brings about a modification in the genes of crops, it affects the biodiversity of a whole range of crops. This increases the risk of gene pollution, as the genes of the modified crops might flow to the wild varieties of other crops through the spread of pollen. With the increas-ing use of GM varieties of seeds, there is a possibility that the domestic varieties of seeds would become extinct. With the arrival of Bt cotton seeds on the market, it has already become difficult to find domestic varieties of seeds. Once all these problems start, according to Mr Mishra, it would become impossible to repair the damage brought about by GM technology. In the case of technologies like cell phones or television, if something goes wrong, one can always go back to the older version of those technologies. But the same cannot be done for GM technology. For example, once Bollgard-II comes to the market replacing Bollgard-I, it becomes difficult and impossible to go back in time to Bollgard-I, or any domestic seed varieties.

Scientists have been arguing that farmers always opt for better varieties of seeds to enhance the quantity and quality of their agricultural crops. So if GM seeds are better, then farmers would definitely choose those seeds. They have argued that if non-Bt seeds are not available in seed shops now, it is because nobody wants them. Farmers prefer to purchase Bt seeds because they give higher production than non-Bt seeds, and shopkeepers are not ready to sell non-Bt seeds because of lack of demand. Scientists supporting GM technology have argued that farmers know very well what is good for them.

Mr Mishra counters this argument by saying that:

> Today, we all know how harmful has been the use of chemical pesticides for growing crops. There is a place in Punjab where a particular train is named as 'cancer train' because it carries cancer patients. Most of these patients belong to the village where chemical pesticides were sprayed in heavy amount to grow crops. The consequence is that, after consuming these pesticide-sprayed crops, people have become sick and got the threatening disease in the form of cancer. Despite knowing all this harmful effect of chemical pesticide spraying, still farmers have been spraying them in their farms till today. Their intention is only to increase crop production and thus maximise their profits. But they are least bothered about the harmful consequences of either chemical pesticides or GM technology. So farmers are naïve and therefore it is a wrong approach to rely on farmers for everything and argue that it is their choice to buy a particular kind of seeds. After all, those crops in the form of GM foods will be received by consumers. So it is also the right of consumers to know what they are eating. We are talking about all these field trials, but we don't have a proper labelling mechanism in place. If a consumer is buying a particular food, he or she has the right to know whether that food is GM or non-GM. So where is transparency? The whole issue needs to be analysed from the point of biodiversity and health, whether toxins of GM has any harmful impact on the health of animals and humans.

The major point Mr Mishra is making here is that farmers are also part of the system and they think in terms of earning money and making profits rather than thinking about biodiversity and health issues. So it would be a wrong approach to rely completely on them rather than create transparency and accountability into the system. The vulnerable condition of the farmer can be understood by the fact that despite knowing the harmful consequences of using chemical pesticides, farmers have been increasingly using them even today. Farmers are only one set of consumers. There are other consumers in the food market as well. Mr Mishra is making the point here that before granting approval to any GM food crops, a system needs to be developed so that consumers can explicitly make a choice on whether to have GM or non-GM food. Mr Mishra has also pointed out in his argument that it is a proven fact that Bt cotton seeds have been a failure and that they cannot help in increasing cotton production. He makes his point by saying:

> GM technology cannot increase the production in agriculture. First of all, remove this GM technology and then let us search for other technologies. We have certain indigenous technologies, which can help in the growth of production. But let us first make it very clear that GM technology has failed in delivering higher agricultural production and then in a fresh manner, let us look for other available technologies, which can really help farmers in increasing their food production.

As the representative of BKS, Mr Mishra wants to discard GM technology completely, as he believes that it has brought no good to Indian society. It should be discarded first and then other alternatives to enhance the quality and quantity of food can be thought of, he says.

However, scientists from regulatory bodies have alleged that many self-proclaimed independent scientists from civil society groups have never made any field visits and never done plant breeding in their lives and yet they proclaim themselves to be independent. One of the scientists from GEAC said:

> As far as government is concerned, we are going for a policy case by case. Even if any GM crop has been approved by America, Brazil or Argentina, we are not concerned. We are going to study case by case before approving any technology. Now definitely the industry will want to promote their technology. Nobody is going to do business in India or anywhere else for charity. So that is very well understood and so there are oppositions too. Oppositions also have their own agenda like, to promote their organic food crops, chemical pesticides or they may be just pure environmental activists. Some of them may not even understand the technology, some of them may understand the technology but ideologically, they might have a different stand. So there are several factors on why globally the whole issue is divided. We can't pinpoint

one reason. But as far as government of India is concerned, I think we take a neutral position. We take all the facts into consideration because it should be science based and need based. That is how we go about it. We are not concerned why the other person is saying yes or no. We definitely look into everything which is being told to us and then examine and then take case by case scientific facts to analyse if it is need based or not.

The GEAC member scientist argues that all the stakeholders have their own interest in supporting or opposing the policy, which is still in the process of development. But the government regulatory bodies are fully committed to taking a neutral decision by finding the facts and steering clear of the influence of the stakeholders of GM technology, he says.

Conclusion

After analysing the arguments of all the interviewees, a lot of criticism and counter-criticism is taking place on both the sides. It has been observed that most of the scientists working in government institutes or holding positions in the regulatory bodies have been supporting the use of GM technology and want field trials for other crops approved. They feel that denying or delaying permission to conduct field trials by the government is like putting restrictions on the independent conduct of science. According to them, only if scientists are allowed to do field trials will they be in a position to analyse the utility of the technology.

On the contrary, several scientists working for NGOs or running their own NGO, mobilise other civil society groups to protest the decisions of the government whenever it tries to approve field trials for GM crops. They argue that India is not ready to conduct open field trials on GM crops, as the suitable infrastructure for that is not in place. They point out that up to now, adequate guidelines for a labelling mechanism has not been developed by the regulatory bodies to distinguish between GM and non-GM food products. They have criticised scientists at the regulatory bodies for not taking into consideration that pollen can travel long distances and therefore ignoring the threat that pollen from GM crops might contaminate wild or domestic varieties of other crops, which will in turn affect biodiversity. They have also blamed scientists for not taking seriously health-related issues. Dr Sagari R. Ramdas and her NGO Anthara have shown in their case history report how some of the animals have been falling sick after eating Bt cotton plants; they have demanded a thorough investigation into the matter from the country's reputed veterinary labs. Other scientists like Dr Bhargava have demanded that chronic rather than acute toxicity tests be conducted during the complete life span of the animals consuming GM crops to determine health safety. Some of the civil society group members have been blaming some scientists who they believe share a common interest with the private companies making these GM technologies, as the scientists themselves receive substantial benefits from them. These groups claim that such scientists want to conduct field trials without a sufficient infrastructure in place only to speed up commercialising these GM crops.

However, most of the scientists at the government institutes and regulatory bodies have rejected these criticisms by civil society members and likened their position to skating on thin ice. They have argued that they have done sufficient tests to know how good GM technology can be for agriculture before arriving at the decision to go for open field trials. They have called the scientists working for civil society groups 'scientists for passion', when they should have been 'scientists for reason'. They proclaim themselves to be independent scientists but have never visited field sites and so do not understand field realities. They have accused the civil society group members opposing GM technology of having their own hidden agenda of promoting their organic crops or chemical pesticides.

Whatever the hidden agendas of the sides supporting or opposing GM technology, from the analysis of their arguments, it becomes clear that the policy elites supporting GM crops give utmost priority to improving agricultural production and farmers' economic status. Similarly, the discourse by policy elites opposing GM technology makes it clear that their priority is environmental friendliness, health safety, freedom of choice and farmers' well-being in the long run.

From this analysis, it can be seen that various interests are shaping the views of the members of different organisations. The majority of government institute scientists have been supporting the use of GM technology for agricultural purposes and they are more in number in the regulatory bodies. There are also scientists working for the civil society organisations and many of them oppose the use of GM technology. It has been observed that whenever the regulatory bodies, where the pro-GM technology scientists are more, gave approval for field trials of GM crops like Bt brinjal, members of civil society groups succeeded in organising themselves and holding massive protests against the approval of the regulatory bodies. As far as bureaucracies and ministries in the government are concerned, they need to take a political decision on the matter. However, their decisions have been highly influenced by both sides. For example, when the Environment Ministry under Jairam Ramesh was set to give consent for Bt brinjal field trials in the open environment, massive protests by civil society groups erupted and the ministry was forced to order a temporary moratorium on field trials. However, corporate groups formed their own pro-GM lobbies to oppose the moratorium by the government. As a result, the prime minister reshuffled the ministry, and later the new ministry granted approval for confined field trials[20] for a few GM crops. After the approval from the central government, the next task for the pro-GM lobbies was to influence the decisions of the state governments in their favour, as agriculture is a state subject and state approval is essential before conducting the trials. As a result, some of the states like Maharashtra, Punjab, Haryana and Andhra Pradesh have given a no-objection certificate (NOC) for the trials, while Madhya Pradesh and Rajasthan, among others, have banned the test.[21] Most of the states have still not taken any decision on the matter. In this way, both pro- and anti-GM lobbies are trying their best to influence government bodies in their political decisions.

As far as political parties are concerned, their view on GM technology keeps fluctuating depending on whether they are in the government or in coalition with other parties. In the case of the Bharatiya Janata Party (BJP), now in power, it is facing opposition from its own party wings like SJM and BKS, which are part of Rashtriya Swayamsevak Sangh (RSS), whenever it tries to approve GM crops for field trials.

Similar rifts can be seen among the left political parties as well. The Left Democratic Front (LDF), a coalition of left parties ruling Kerala, has been opposing the use of GM technology in agriculture. However, a rift developed within the alliance when a Marxist Communist Party of India (CPI-M) politburo member, Mr S. Ramachandran Pillai, stated that "GM seeds could be used after ensuring that they clear all the tests and prove that they will not harm the environment and living beings".[22] It is significant to note that Mr Pillai has been the national president of the party's farmers' wing 'Kisan Sabha'. Immediately after Pillai's remark, the then agriculture minister of Kerala, Mr Mullakkara Ratnakaran, said that "there has been no evidence to show that GM crops will benefit the farm sector and their use will wipe out traditional seeds".

One thread common to both the supporters and detractors of GM technologies is how to improve agriculture and the condition of farmers, who are the major stakeholders in the entire issue. Therefore, the next chapter will talk about the priorities and concerns of the farmers cultivating Bt cotton crops in Telangana and Maharashtra.

Notes

1 Though the number of policy elites selected for the interview is small, these elites have been highly influential in the ongoing policymaking processes regarding GM technologies. The interviews were qualitative in nature.
2 Dr Ramesh Sonti was interviewed on 28 October 2014 at his office in CCMB in Hyderabad.
3 Dr Pushpa Mittra Bhargava was interviewed on 24 October 2014 at his office based in Hyderabad.
4 Dr Virender S. Chauhan was interviewed on 10 February 2015 at his office in ICGEB, New Delhi.
5 Dr Vandana Shiva was interviewed on 7 February 2015 at her Navdanya office in New Delhi.
6 Dr Ranjini Warrier was interviewed on 23 February 2015 at her office in Ministry of Environment, Forests and Climate Change, Government of India, New Delhi.

7 Dr K.C. Bansal was interviewed on 24 February 2015 at his NBPGR office in New Delhi.
8 Dr Sagari R. Ramdas was interviewed on 30 October 2014 at her Anthara office in Hyderabad.
9 A farmer in Oregon had found some genetically engineered wheat growing on his land. It was an unwelcome surprise because this type of wheat has never been approved for commercial cultivation. For details please see an article by Dan Charles on "GMO Wheat Found in Oregon Field. How Did It Get There?" Retrieved from www.npr.org/sections/thesalt/2013/05/30/187103955/gmo-wheat-found-in-oregon-field-howd-it-get-there
10 Liberty link rice, which is also known as phosphinothricin-tolerant rice or glufosinat- tolerant rice, is a genetically modified organism developed by Bayer CropScience. This particular rice has been genetically modified to enable it to withstand the spraying of a chemical pesticide called glufosinate. LL rice became the subject of controversy in 2006, when it was found in commercial shipments of long grain rice in the United States, as it had not yet been approved for human consumption.
11 Dr G.V. Ramanjaneyulu was interviewed on 17 October 2014 at his CSA office, Hyderabad.
12 An acute toxicity test involves an assessment of the general toxic effect of a single dose or multiple doses of a chemical or product within 24 hours after its absorption through oral, dermal or inhalation routes and the other effects occurring during the subsequent 21 to 100 days of observation.
13 A chronic toxicity test involves exposure to the test substance by an appropriate route like oral, dermal or inhalation with appropriate dosages for a major portion of the animal's life span.
14 Dr Akhilesh Tyagi was interviewed on 13 February 2015 at his NIPGR office, New Delhi.
15 A CRO (contract research organisation) provides support to the pharmaceutical, biotechnology and medical device industries in the form of research services outsourced on a contract basis. A CRO may provide such services as biopharmaceutical development, biologic assay development, commercialisation, preclinical research, clinical research, clinical trial management and pharmacovigilance. CROs also support foundations, research institutions and universities in addition to governmental organisations.
16 Dr Ramesh Chand was interviewed on 5 February 2015 at his NIAP office, New Delhi.
17 Dr Suman Sahai was interviewed on 3 February 2015 at her Gene Campaign office, New Delhi.
18 Dr Aswani Mahajan was interviewed on 11 February 2015 at his SJM office based in New Delhi.
19 Mr Mohini Mohan Mishra was interviewed on 24 February 2015 at his BKS office based in New Delhi.
20 See the *Live Mint* newspaper dated 28 February 2014 in the 'Politics' section. Neha Sethi reports in the copy titled "Veerappa Moily Approves Trials of Genetically Modified Crops". Retrieved February 6, 2017, from www.livemint.com/Politics/xhXZNcGHVDq0wRz0ofR4CI/Veerappa-Moily-approves-trials-of-genetically-modified-crop.html
21 See the *Business Line* newspaper dated 30 January 2015 in the 'news science' section. The copy is titled "Maharashtra Approves Field Trials for Four GM Crops". Retrieved February 6, 2017, from www.thehindubusinessline.com/news/national/maharashtra-approves-field-trials-for-four-gm-crops/article6839527.ece

22 See the *DNA: Daily News & Analysis* newspaper dated 2 January 2011 in the 'India' news section. The copy is titled "CPI-M Leader Backs Use of Genetically Modified Seeds". Retrieved February 6, 2017, from www.dnaindia. com/india/report-cpi-m-leader-backs-use-of-genetically-modified-seeds-1489185

6 What do farmers want?

Field survey results of Telangana and Maharashtra

Introduction

After highlighting the first phase of the field study in the previous chapter where mostly interviews with policy elites were discussed, this chapter deals with the second and third phases of the field study, where mostly the cotton-growing farmers from Telangana and Maharashtra were interviewed. A closed-ended questionnaire was prepared to talk to 100 cotton-growing farmers each from Telangana and Maharashtra. Within the states of Telangana and Maharashtra, I chose Warangal and Yavatmal districts, respectively, for the field survey. Farmers in these two regions are highly dependent on cotton cultivation for their livelihood. Reports of them committing suicide have been increasing specifically from these two regions due to losses faced by them in agriculture. Before I discuss the perspectives of farmers regarding Bt cotton cultivation in both regions, it will be significant to draw a brief outline of both districts separately.

Warangal district of Telangana

In Warangal, 42 per cent of the area is under cultivation, and agriculture provides employment to 69 per cent of the workforce (Human Development in Telangana State District Profiles, 2015). Some of the important crops grown here are rice, maize, green gram, cotton and chillies. The district has 78 per cent of its area under irrigation through ground water. Cotton is an important crop here and accounts for 37 per cent of gross cropped area (GCA). After cotton, rice is another important crop, which accounts for 34 per cent of GCA. It is significant to note that agriculture contributes 21 per cent of gross district domestic product (GDDP), while 69 per cent of the workforce depends on agriculture. The share of agriculture in employment increased by 1.3 percentage points, while its share in GDDP decreased by 6.2 percentage points. The share of industry in GDDP is 21 per cent, while its share in employment is 10 per cent. More than 50 per cent of the GDDP comes from the services sector, but it employs only 22 per cent of the workforce. Therefore, though the share of agriculture in GDDP is very low, 69 per cent of the work force is dependent on it. This shows the significance of agriculture in the state and the crisis it is going through at the moment. In this context, it would be interesting to see the adoption of new genetically modified technology in the form of Bt cotton by farmers in the hope of increasing their yield and net income. Bt cotton was brought to the market with the promise that it would help farmers in reducing their expenditure on chemical pesticides and in protecting the cotton plants from harmful insects called bollworms, which have been causing extensive damage to the cotton plants. Thus, if Bt as a technology provides protection to the plants from harmful bollworms, there would automatically be an increase in the cotton yield. Therefore, it would be interesting to know the experiences of the farmers on whether Bt cotton seeds helped in serving this purpose.

Yavatmal district of Maharashtra

Yavatmal district falls in Vidarbha region, and it is here that the number of farmers committing suicide has been the highest in India. Yavatmal is one of the 35 districts in Maharashtra state in the Indian subcontinent. It is located in the east-central part of the state. According to the 2011 census, the total population of this district is 2,772,348, including a tribal population of 469,000. The tribal background of farmers in this district and their low educational attainment have kept most of them economically 'backward' and socio-culturally 'traditional' (Talule, 2013). Also, the low level of industrialisation and inadequacy of logistic facilities across the entire Vidarbha region have kept these people confined to agriculture for their livelihood. In Yavatmal, 74.75 per cent of the total geographic area is under cultivation, of which 37.52 per cent is under food grains and 44.74 per cent under cotton, followed by pulses with 19.70 per cent, oil seeds with 15.40 per cent, jowar with 14.10 per cent, wheat 2.49 per cent and sugarcane 1.24 per cent.

Table 6.1 shows the average yield of some crops in kg per hectare grown in Yavatmal district of Maharashtra.

Table 6.1 Yield per hectare of crops

Crops	Per hectare yield in kg
Jowar	1082
Mung	577
Wheat	1803
Gram	848
Urad	399
Cotton	412

Source: Talule (2013)

The gross irrigated area of the district is 60,520, whereas the net irrigated area is 45,958 hectares. This is about 5.12 per cent of the total cultivable area of the district. Crop-wise the irrigated areas are food crops – 45.54 per cent, cereals – 15.80 per cent, sugarcane – 18.52 per cent, fruits and vegetables – 5.68 per cent, cotton – 1.11 per cent and oil seeds – 6.69 per cent (Talule, 2013).

Most of the farmers in the district grow cotton crops. In recent times, cotton farming has been suffering a lot due to irregular rainfall. As shown earlier, though cotton is grown in 44.74 per cent of the area, only 1.11 per cent of the cotton-growing area is under irrigation. The rest of the area is rain-fed and depends on monsoons. Farmers adopted Bt cotton seeds in the hope that these seeds would help them in reducing the spraying of chemical pesticides, thus reducing their expenditure and protecting cotton plants from boll-worms. In this context, it would be interesting to see how far Bt cotton seeds have been successful in Maharashtra.

Field survey of farmers in Telangana and Maharashtra

To understand the perception of farmers on Bt cotton, 100 cotton-growing farmers were interviewed in Warangal district of the Telangana region. Within Warangal district, two blocks have been selected for the field survey: Hasanparthi and Atmakur. The total size of the sample is 100 farmers from the two blocks taken together. From each block, 50 farmers have been selected for interviews with the help of a closed-ended questionnaire. Within Hasanparthi block, I visited two villages, viz. Mucharla and Ganturpalli. From each of these two villages, I interviewed 25 farmers through snowball sampling[1] methodology for the field survey. Within Atmakur block, I visited three villages, viz. Urugonda, Gudapadu and Atmakur. In Urugonda 14 farmers were chosen, in Gudapadu 17 farmers were chosen and in Atmakur 19 farmers were chosen for interviews through snowball sampling methodology.

Similarly, within the district of Yavatmal in Maharashtra, two blocks were selected for the field survey: Ghatanji and Pandharkawda. The total size of the sample was 100 with both the blocks taken together. From each block, 50 cotton-growing farmers were selected to answer a closed-ended questionnaire. Within each block, four villages were selected and within each village, 12 to 13 farmers were chosen through snowball sampling methodology for the survey.

The questionnaire was prepared to look for answers from farmers of both Telangana and Maharashtra on the impact of cultivation of Bt cotton on yields, returns, seed usage and expenditure, pesticide usage and expenditure, pest attacks, availability of irrigation facilities, reasons for farmer suicides, human health, livestock health, soil quality and impact on the environment.

Categorisation of farmers

On the basis of land size,[2] farmers in both the regions of Telangana and Maharashtra have been classified into five farming sections: large, medium, semi-medium, small and marginal farmers. Therefore, analysing the types of farmers on the basis of the Agriculture Census 2010–11 (Census, 2010–11), the number of large farmers in Telangana was found to be 0, the number of medium farmers was found to be 1,[3] the number of semi-medium farmers was found to be 33, the number of small farmers 36 and the number of marginal farmers 30.

Similarly, in Maharashtra, the number of large farmers was found to be 2, the number of medium farmers was found to be 18, the number of semi-medium farmers was found to be 42, the number of small farmers was found to be 33 and the number of marginal farmers was found to be 5. After grouping the farmers of both regions on the basis of their land holding into five sections of large, medium, semi-medium, small and marginal, we found interesting results regarding their status on class and caste groupings, as presented in Table 6.2.

Farmers cultivating cotton as the main crop

Since the field survey mainly deals with the cotton-growing farmers, it was important for us to ask the farmers whether they grow cotton as their main crop or secondary crop. In Table 6.3, the number of farmers who grow cotton as the main crop is given. During the field survey, it was observed that

Table 6.2 Castes of farmers

State	Farmer type	General	OBC	SC/ST	Total
Telangana	Large	0	0	0	0
	Medium	1	0	0	1
	Semi-medium	24	9	0	33
	Small	14	14	8	36
	Marginal	4	19	7	30
	Total	43	42	15	100
Maharashtra	Large	0	1	1	2
	Medium	2	5	11	18
	Semi-medium	6	13	23	42
	Small	4	12	17	33
	Marginal	0	2	3	5
	Total	12	33	55	100

Source: Primary field survey

Table 6.3 Farmers growing cotton as the main crop

State	Farmer type	Number of farmers growing cotton as the main crop	Number of farmers growing cotton as a secondary crop	Total
Telangana	Large	0	0	0
	Medium	1	0	1
	Semi-medium	32	1	33
	Small	36	0	36
	Marginal	29	1	30
	Total	98	2	100
Maharashtra	Large	2	0	2
	Medium	18	0	18
	Semi-medium	42	0	42
	Small	33	0	33
	Marginal	5	0	5
	Total	100	0	100

Source: Primary field survey

farmers who had a large area under cotton responded that cotton was their most important crop.

From table 6.3, it is clear that the majority of farmers in both Telangana and Maharashtra have been cultivating cotton as their main crop. In Telangana, it was observed that during the entire Kharif season,[4] the main crop for the farmers in the Warangal district of Telangana has been cotton. However, during the Rabi season,[5] they switch over to the cultivation of maize in the irrigated areas and in non-irrigated areas they may cultivate bajra.

In Maharashtra, all farmers responded that cotton is their main crop and they cultivate it throughout the year. Along with cotton, the other crops they cultivate are mostly toor (pigeon peas), jowar (sorghum) and soya bean. Some farmers also grew wheat in place of soya bean as a secondary crop. The reason they grow cotton as their primary crop and others such as toor, jowar, soya bean and wheat as secondary crops is that cotton plants keep flowering throughout the season whenever they are irrigated or receive rain, but for the other crops farmers have to wait until they are fully grown before harvesting. Therefore, farmers are able to earn from cotton plants more frequently and rapidly in comparison to toor, jowar and soya bean.

Soil quality

Normally black soil is considered good for cotton cultivation. From the survey conducted in the Warangal district of Maharashtra, Atmakur block is rich in black soil while Hasanparthi block is rich with red soil. The fields

Table 6.4 Soil quality

State	Farmer type	Black	Red	Mixed	Total
Telangana	Large	0	0	0	0
	Medium	0	0	1	1
	Semi-medium	12	20	1	33
	Small	18	17	1	36
	Marginal	18	11	1	30
	Total	48	48	4	100
Maharashtra	Large	1	0	1	2
	Medium	9	2	7	18
	Semi-medium	23	8	11	42
	Small	20	5	8	33
	Marginal	2	3	0	5
	Total	55	18	27	100

Source: Primary field survey

of farmers were observed and found to contain either red or black soil or sometimes a mixture of both.

Similarly, from the survey conducted in the Yavatmal district of Maharashtra, soils were found to be mostly black in both the blocks of Ghatanji and Pandharkawda. Still at a few places in both the blocks, the soils were either red or a mixture of black and red. The responses of farmers about the quality of the soil for cultivating cotton have been recorded in Table 6.4.

Suitability of soil to grow other crops

After observing the types of soil, farmers were then asked whether the soil was suitable for growing other crops. The purpose of asking this question was to analyse the perspective of the farmers about the fertility of the soil and whether they give any importance to the cultivation of leguminous crops like peas and pulses to help the soil regain fertility rather than use chemical fertilizers to enrich the soil. Table 6.5 shows the responses of the farmers in this regard.

In Telangana, Table 6.5 shows that out of 100 selected cotton-growing farmers, the majority of them do not give importance to planting leguminous plants for enriching the soil. They simply spray chemical fertilizers to enrich their soil, which may not be good for the health of the soil in the long term. When asked why they do not cultivate leguminous plants, most of them responded that they could not earn much from them, as the price of leguminous crops like pulses fluctuates in the market and no fixed minimum support price has been set by the government for them. Leguminous plants like peas and pulses are not commercial crops, whereas cotton is a

Table 6.5 Suitability of soil

After harvesting the Bt cotton, does soil remain suitable to grow other crops?

State	Farmer type	Yes	No	Can't say	Total
Telangana	Large	0	0	0	0
	Medium	1	0	0	1
	Semi-medium	31	2	0	33
	Small	33	2	1	36
	Marginal	26	1	3	30
	Total	91	5	4	100
Maharashtra	Large	1	1	0	2
	Medium	6	11	1	18
	Semi-medium	25	17	0	42
	Small	11	21	1	33
	Marginal	2	3	0	5
	Total	45	53	2	100

Source: Primary field survey

complete commercial crop. Therefore, in the hope of earning more, most of the farmers have adopted commercial crops like Bt cotton for cultivation. Thus, commercialisation has forced farmers to replace food crops with Bt cotton.

During the field survey, many farmers in Maharashtra said the excessive cultivation of Bt cotton has led to an increase in the spraying of chemical fertilizers. They also said that the weather has drastically changed in the last four to five years. Earlier it used to rain a lot. But now it does not rain adequately. As a result of increased spraying of chemical fertilizers and less rainfall, the soil has started becoming hard and unsuitable for growing other crops. The soil in Maharashtra is mainly losing its fertility because of the excessive use of chemical fertilizer spraying and inadequate irrigation facilities. As a result in Maharashtra, the soil is hard and inappropriate for the cultivation of other crops.

Knowledge of Bt cotton

One of the important objectives during the field survey was to enquire from the cotton-growing farmers whether they can distinguish non-Bt cotton from Bt cotton and whether they are aware of the specific quality of Bt cotton plants – that they kill bollworm insects which attack cotton bolls. Table 6.6 shows their responses in this regard.

From the table, it can be observed that when farmers were asked if they are aware of the term 'Bt' in Bt cotton, they responded differently and

Table 6.6 Knowledge of Bt cotton

Do you know what is Bt in Bt cotton?

State	Farmer type	Yes	No	Total
Telangana	Large	0	0	0
	Medium	1	0	1
	Semi-medium	26	7	33
	Small	20	16	36
	Marginal	13	17	30
	Total	60	40	100
Maharashtra	Large	2	0	2
	Medium	16	2	18
	Semi-medium	30	12	42
	Small	17	16	33
	Marginal	3	2	5
	Total	68	32	100

Source: Primary field survey

their responses were recorded in terms of 'yes' and 'no'. From Table 6.6, it is obvious that most of the farmers growing cotton know the function of Bt in the cotton. However, there were also many farmers who were unable to distinguish Bt from non-Bt cotton.

Collection of information

Since Bt cotton is a new variety of cotton seed developed by biotechnology scientists in their labs, cultivating them requires mechanisms different from conventional seeds. Therefore, it is essential for the farmers to have good knowledge about the cultivation process of Bt cotton. In this regard, it was important to ask farmers about their mechanisms of collecting information regarding Bt cotton cultivation. Table 6.7 shows the responses of the farmers of Telangana and Maharashtra in this regard. Here farmers selected their responses from a set of optional responses in the closes-ended questionnaire.

Farmers' experience growing non-Bt cotton

In the field, it was found that non-Bt cotton seeds were not available either in the market or with any farmers. There can be two varieties of non-Bt cotton seeds. One is the indigenous variety of cotton seed and the other is hybrid varieties. India was famous for its finest fabric woven from fibres of the indigenous Desi cotton. However, before independence, several parts of the country had been evaluated for their suitability for the American cotton

Table 6.7 Collection of information

State	Farmer type	Self-experience	Friends and relatives	Agricultural experts	Seed shopkeeper	Government agency	Kissan call centre	Copying other farmers
Telangana	Large	0	0	0	0	0	0	0
	Medium	1	0	1	0	1	1	0
	Semi-medium	32	13	21	0	7	13	0
	Small	34	10	23	0	7	13	0
	Marginal	29	10	15	1	2	8	0
	Total	96	33	60	1	17	35	0
Maharashtra	Large	2	2	0	0	0	2	1
	Medium	17	6	7	0	3	13	2
	Semi-medium	41	26	12	0	2	32	6
	Small	32	23	10	0	5	26	4
	Marginal	5	4	1	0	0	3	0
	Total	97	61	30	0	10	76	13

Source: Primary field survey

species, *Gossypium hirsutum*, which is especially suited for mechanized spinning mills. After independence, when the spinning mills suffered from the lack of raw fibres of American cotton, efforts were intensified to identify regions suited for the long-staple American varieties.

Hybrid cotton was considered a way to obtain high yields. But their full potential could be harnessed only under optimal conditions, i.e. availability of adequate irrigation facilities, and pesticides and fertilizers are significant for the cultivation of hybrid varieties of cotton.

Now to compare the yield of Bt cotton seeds with non-Bt seeds was impossible, as no counterfactual effect was available. To deal with this problem, farmers selected for the survey were asked whether they had experience growing non-Bt seeds before Bt seeds came on to the market. Table 6.8 presents the number of farmers experienced in non-Bt cotton cultivation.

Therefore, from this table, it is clear that there were significant number of farmers both in Telangana and Maharashtra who had experience growing non-Bt cotton crops.

Comparison of Bt cotton seeds with non-Bt cotton seeds

The farmers who had experience growing Bt as well as non-Bt cotton crops were then asked to compare yields and say whether Bt or non-Bt cotton seeds give more yield. Table 6.9 shows the responses of farmers in this regard. Those farmers who did not have experience in cultivating non-Bt cotton crops were not considered in this question.

Table 6.8 Farmers who have experience growing non-Bt cotton

State	Farmer type	Experienced in non-Bt cotton cultivation	Not experienced in non-Bt cotton cultivation	Total
Telangana	Large	0	0	0
	Medium	1	0	1
	Semi-medium	24	9	33
	Small	29	7	36
	Marginal	22	8	30
	Total	76	24	100
Maharashtra	Large	1	1	2
	Medium	13	5	18
	Semi-medium	29	13	42
	Small	24	9	33
	Marginal	2	3	5
	Total	69	31	100

Source: Primary field survey

Table 6.9 Yield comparison of Bt cotton with non-Bt cotton

State	Farmer type	More yield in Bt cotton	More yield in non-Bt cotton	Equal yield	N.A.	Total
Telangana	Large	0	0	0	0	0
	Medium	1	0	0	0	1
	Semi-medium	21	2	1	9	33
	Small	27	1	1	7	36
	Marginal	20	1	1	8	30
	Total	69	4	3	24	100
Maharashtra	Large	1	0	0	1	2
	Medium	11	2	0	5	18
	Semi-medium	27	2	0	13	42
	Small	15	6	3	9	33
	Marginal	2	0	0	3	5
	Total	56	10	3	31	100

Source: Primary field survey

In this table, it can be observed that most of the farmers who had experience cultivating non-Bt cotton responded that adoption of Bt cotton seeds helped them in increasing the cotton yield. Also, according to the official government report (Directorate of Economics and Statistics, 2012), after the introduction of Bt cotton, cotton production in the country has increased immensely. However, some civil society organisations like Greenpeace, Gene Campaign, Navdanya and members of coalition called ASHA and Coalition for GM-Free-India have been arguing that Bt cotton helps in protecting cotton against the bollworm only for few years and after that the bollworm becomes resistant to the Bt cotton and thereafter Bt cotton is ineffective to the bollworms. As a result, farmers have to spray more chemical pesticides to protect the cotton, which ultimately increases the total input costs of farmers in growing cotton. Therefore, though Bt cotton cultivation helps farmers increase the yield, this is only for a certain period. Thereafter, the technology becomes outdated and needs to be replaced by a newer variety or newer generation of Bt cotton (Navneet, 2014).

Bt cotton was adopted by Indian farmers mainly because it helped them protect the cotton plants from bollworms. The Desi (local) cotton species was generally sturdy and highly resistant to almost all biotic and abiotic stresses. But they were in general susceptible to pink bollworm, *Pectinophora gossypiella*, and the spotted bollworm, *Earias vitella*. They were resistant to most of the other pests. But the Desi varieties of cotton were not suitable for the spinning mills, as they had a short staple. For spinning mills, growing of long staple American cotton, *G. hirsutum* was required. As a result, in India, efforts were intensified to identify the regions where long-staple hybrid varieties of American cotton could be cultivated. But these hybrid varieties of cotton were susceptible to the bollworms. To protect the hybrid cotton crops from bollworms, pesticide spraying increased. Bt cotton was introduced to reduce the use of pesticides.

Cost comparison of Bt cotton seeds with non-Bt seeds

Based on the experience of farmers cultivating Bt and non-Bt cotton, they were asked to compare the cost prices of Bt seeds with that of non-Bt seeds. The responses of the farmers have been captured in Table 6.10.

In Table 6.10, it can be observed that the responses of farmers have been recorded in both the states of Telangana and Maharashtra in the form of N.A. (not applicable), 'yes' and 'no'. The farmers who did not have experience cultivating non-Bt crops were given N.A. for the response. A response of 'yes' signifies Bt crops are costlier than non-Bt, while a response of 'no' signifies non-Bt crops are costlier than Bt crops.

Therefore, in both the states of Telangana and Maharashtra, among the farmers who were interviewed, every farmer who had experience cultivating non-Bt cotton crops responded that Bt cotton is costlier than non-Bt crops. However, it was expected that with the adoption of Bt cotton seeds by farmers, the overall cultivation of Bt cotton would become cheaper than the cultivation of non-Bt cotton crops, especially the hybrid variety of seeds. This is because farmers had to spend lot on the spraying of pesticides and fertilizers

Table 6.10 Cost comparison of Bt cotton seeds with non-Bt seeds

Are Bt cotton seeds comparatively costlier than non-Bt cotton seeds?

State	Farmer type	N.A.	Yes	No	Total
Telangana	Large	0	0	0	0
	Medium	0	1	0	1
	Semi-medium	9	24	0	33
	Small	7	29	0	36
	Marginal	8	22	0	30
	Total	24	76	0	100
Maharashtra	Large	1	1	0	2
	Medium	3	15	0	18
	Semi-medium	11	31	0	42
	Small	9	24	0	33
	Marginal	2	3	0	5
	Total	26	74	0	100

Source: Primary field survey

on the hybrid varieties of cotton seeds. This used to increase the overall expenditure of farmers on cotton crops. Bt cotton seeds were adopted by farmers because initially they helped them in reducing the spraying of chemical pesticides, thereby reducing the overall expenditure of farmers on cotton cultivation for a limited period.

Further, most of the farmers in both the states responded that they have to take out loans for Bt cotton cultivation. They had to take out loans also when they used to cultivate the hybrid variety of cotton crops. Adoption of Bt cotton seeds did help to a certain extent in reducing the expenditure on pesticides but not completely. After time, a number of other insects started growing and attacking cotton plants in the absence of bollworms. This compelled farmers again to increase pesticide spraying and as a result the expenditure on cotton crops again increased. To meet the expenditure, farmers have to take out loans from the government or private banks.

Table 6.11 shows the responses of the farmers in terms of taking out a loan to purchase Bt cotton seeds.

The cotton-growing farmers were further asked whether Bt cotton cultivation helped them in increasing their net income to clear their debts. Their responses in this regard have been captured in Table 6.12.

In Table 6.12, the responses of the farmers have been recorded in the form of N.A., 'yes' and 'no'. Those farmers who did not have experience cultivating non-Bt cotton crops have been labelled N.A. A response of 'yes' indicates that Bt cotton cultivation benefited them in raising their incomes, and a 'no' response indicates Bt cotton cultivation did not help them in raising their income.

Table 6.11 Taking out a loan to purchase Bt cotton seeds

State	Farmer type	Yes	No	Total
Telangana	Large	0	0	0
	Medium	0	1	1
	Semi-medium	25	8	33
	Small	24	12	36
	Marginal	11	19	30
	Total	60	40	100
Maharashtra	Large	2	0	2
	Medium	15	3	18
	Semi-medium	38	4	42
	Small	32	1	33
	Marginal	2	3	5
	Total	89	11	100

Source: Primary field survey

Table 6.12 Whether Bt cotton cultivation helped in increasing the net income so that debt can be cleared

State	Farmer type	N.A.	Yes	No	Total
Telangana	Large	0	0	0	0
	Medium	1	0	0	1
	Semi-medium	8	21	4	33
	Small	12	23	1	36
	Marginal	19	10	1	30
	Total	40	54	6	100
Maharashtra	Large	0	1	1	2
	Medium	3	14	1	18
	Semi-medium	4	34	4	42
	Small	1	18	14	33
	Marginal	3	2	0	5
	Total	11	69	20	100

Source: Primary field survey

Therefore, this table shows that most of the farmers selected for the field survey responded that the cultivation of Bt cotton did help them in increasing their net income because of an increase in cotton yield. However, they often also suffered losses in the cultivation of cotton crops, but this was because of low rainfall and inadequate irrigation facilities and not because of the use of Bt cotton seeds.

Reasons for farmers committing suicide

To understand the main reason for farmers growing Bt cotton committing suicide, farmers who were interviewed were asked if they knew any cotton-growing farmers in the village who attempted suicide or had committed suicide. Thereafter, they were asked whether they knew why the farmers resorted to such action. The responses of most of the farmers on this matter were their inability to clear debt. They took out a loan either from the national or state banks or from some private agencies to plant cotton. But any failure of a crop due to a lack of rain and irrigation facilities compelled them to take such extreme steps. Tables 6.13 and 6.14 throw more light on this regard.

From Table 6.13, it can be observed that most of the farmers know cotton-growing farmers who committed suicide. Therefore, those farmers who responded 'yes' in Table 6.13 were further asked about the possible reasons for the farmers committing suicide. Table 6.14 shows the responses of the farmers in this regard.

In Table 6.14, the responses of the farmers have been recorded in the form of N.A., 'debt', and 'other reasons'. The farmers who responded 'no' in Table 6.13

Table 6.13 Enquiring about farmers committing suicides

Has there been any cotton-growing farmer in your village who committed suicide?

State	Farmer Type	Yes	No	Total
Telangana	Large	0	0	0
	Medium	1	0	1
	Semi-medium	31	2	33
	Small	36	0	36
	Marginal	27	3	30
	Total	95	5	100
Maharashtra	Large	1	1	2
	Medium	15	3	18
	Semi-medium	38	4	42
	Small	32	1	33
	Marginal	3	2	5
	Total	89	11	100

Source: Primary field survey

Table 6.14 Reasons for farmers committing suicides

What is the reason for farmers committing suicide?

State	Farmer Type	N.A.	Debt	Other Reasons	Total
Telangana	Large	0	0	0	0
	Medium	0	1	0	1
	Semi-medium	1	32	0	33
	Small	0	36	0	36
	Marginal	2	28	0	30
	Total	3	97	0	100
Maharashtra	Large	1	1	0	2
	Medium	3	14	1	18
	Semi-medium	4	36	2	42
	Small	1	30	2	33
	Marginal	2	3	0	5
	Total	11	84	5	100

Source: Primary field survey

have been considered not applicable for a response in Table 6.14, as they do not know farmers who had committed suicide.

Most of the farmers in Table 6.14 responded that being under a huge debt was the major reason for farmers committing suicide in both states. Most of these farmers went into debt because of the failure of the Bt cotton cultivation. The reason for this failure has been the scarcity of rainfall and the unavailability of adequate irrigation facilities. According to the study conducted by Thomas and Tavernier (2017), there is no dispute about the utility and viability of the Bt cotton, but the national and interstate data do not support transgenic cotton as an overall boon to the Indian farmers. The unfavourable expenditure–yield ratio can place a Bt farmer in a more precarious position than a non-Bt farmer. According to Thomas and Tavernier, the accusations of biotech opponents regarding patent imposition and usage of terminator technology can be baseless, but there is a definite association between economic factors associated with Bt cotton farming and farmer suicides. It should be kept in mind that by the time India adopted Bt cotton, hybrid cultivation had already proven detrimental and provided fodder for the critics (Thomas & Tavernier, 2017).

Experience of failure in Bt cotton cultivation

Failure to pay back the debt to the government or private banks was the major reason for most of the suicides among farmers. The reason for their failure to pay the debt is insufficient yield from Bt cotton cultivation. The cotton growers in the region are highly dependent on monsoon rains for the growth of cotton plants. The weather has changed in the last few years drastically and monsoon rains have not been coming on time. As a result, the cotton plants dry up in the absence of a proper irrigation system. Those farmers who have a well can provide irrigation to their plants for some months. But the wells also go dry because of extreme hot weather and overuse of ground water.

During the field survey, the farmers were asked whether they ever faced any failure during cotton cultivation in terms of good yield. 'Failure' here is defined as failure of cotton cultivation due to scarcity of water and not because of the use of new seed technology. Table 6.15 shows their responses in this regard.

Reasons for the failure of Bt cotton cultivation

In Telangana, all the farmers who said that they faced failure in Bt cotton cultivation responded that low rainfall and inadequate irrigation facilities were the main reason. Table 6.16 shows the responses of only Telangana farmers. The responses of Maharashtra farmers on this question have been shown separately in Table 6.17 with a slight modification in the question.

Farmers who responded that they have not experienced failure in cotton cultivation in Table 6.15 have been considered not applicable for a response in Table 6.16. Among the farmers who responded that they experienced failure, all of them cited inadequate irrigation facilities and low rainfall as the reason for the crop failure.

Table 6.15 Experiencing failure in Bt cotton cultivation

Have you ever experienced failure in Bt cotton cultivation?

State	Farmer Type	Yes	No	Total
Telangana	Large	0	0	0
	Medium	0	1	1
	Semi-medium	24	9	33
	Small	32	4	36
	Marginal	27	3	30
	Total	83	17	100
Maharashtra	Large	2	0	2
	Medium	15	3	18
	Semi-medium	35	7	42
	Small	30	3	33
	Marginal	4	1	5
	Total	86	14	100

Source: Primary field survey

Table 6.16 Responses of Telangana farmers regarding cotton failure due to low rainfall and inadequate irrigation facilities

Is cotton failure due to low rainfall and inadequate irrigation facilities?

State	Farmer Type	N.A.	Yes	No	Total
Telangana	Large	0	0	0	0
	Medium	1	0	0	1
	Semi-medium	9	24	0	33
	Small	4	32	0	36
	Marginal	3	27	0	30
	Total	17	83	0	100

Source: Primary field survey

Table 6.17 Reasons for the failure of Bt cotton cultivation in Maharashtra

State	Farmer Type	N.A.	Due to Low Rainfall	Due to Inadequate Irrigation Facilities	Due to Presence of Counterfeit Seeds
Maharashtra	Large	0	2	2	2
	Medium	3	15	15	14
	Semi-medium	7	35	35	35
	Small	3	30	30	30
	Marginal	1	4	4	4
	Total	14	86	86	85

Source: Primary field survey

Similarly, Maharashtra farmers were asked to choose their responses regarding the reason for crop failure from a set of options. Table 6.17 shows their responses.

Though farmers in Maharashtra agreed that all three factors – lack of rainfall, scarcity of water for irrigation and the presence of counterfeit seeds in Bt cotton packets – were responsible for the failures in cotton cultivation, they stressed that a lack of rainfall and inadequate irrigation facilities were the main culprits.

Kinds of irrigation facilities

From the Table 6.17, it is obvious that low rainfall and inadequate irrigation facilities are responsible for cotton crop failures. In this context, it was considered important to find out from farmers their sources for irrigation. Table 6.18 shows the responses of farmers on the sources they use.

In Table 6.18, it can be observed that majority of farmers in Telangana and Maharashtra are either dependent on wells or monsoon rains for irrigation. Only a few farmers have other techniques like a drip or lift system and bore pump for irrigation. Therefore, when monsoons are delayed and wells dry up because of the excessive use of groundwater, farmers face difficulties in cotton cultivation.

Guidelines for Bt cotton cultivation

The purpose of introducing Bt cotton was to reduce the spraying of chemical pesticides, which was harmful for the health of farmers. Chemical pesticides were sprayed to kill the insects that harmed cotton crops. The

Table 6.18 Kinds of irrigation facilities

State	Farmer Type	Well	Drip/ Lift	Bore Pump	Monsoon	Lake	Canal	Total
Telangana	Large	0	0	0	0	0	0	0
	Medium	0	0	1	0	0	0	1
	Semi-medium	23	5	3	1	0	1	33
	Small	25	1	0	9	1	0	36
	Marginal	24	0	0	6	0	0	30
	Total	72	6	4	16	1	1	100
Maharashtra	Large	1	0	0	1	0	0	2
	Medium	7	2	4	3	0	2	18
	Semi-medium	13	4	2	21	0	2	42
	Small	8	5	1	18	0	1	33
	Marginal	1	0	0	4	0	0	5
	Total	30	11	7	47	0	5	100

Source: Primary field survey

bollworm used to be the major insect causing maximum damage to the plant in comparison to other insects. To control bollworms, Bt cotton was developed by biotechnology scientists in the lab. It was assumed that if cotton plants were protected from bollworm insects, cotton production would automatically increase. Initially when Bt cotton was approved for commercial cultivation, it did help farmers in reducing the spraying of chemical pesticides, but after time, the spraying of chemical pesticides increased again. The reason was that eventually the bollworm insects developed immunity to Bt cotton. Thus the Bt cotton plants, which were able to produce their own insecticide to kill the bollworms attacking them, became ineffective against them in the long run. However, scientists came up with a temporary solution for this. They advised farmers to adopt the practice of refugia to delay the development of immunity in bollworm insects. To practice refugia, farmers would have to cultivate non-Bt cotton crops in a small area of land around Bt cotton plants and that area must not be sprayed with chemical pesticides. As a result, the normal bollworm insects that attack the non-Bt cotton plants can mate with those bollworm insects that have become resistant to the Bt cotton plants, thereby producing offspring which would be no longer be resistant to the Bt cotton plants.

However, during the field survey in the Warangal and Yavatmal districts of Telangana and Maharashtra, respectively, it was observed that either the cotton-growing farmers are unaware of the practice of refugia, or if they are aware, they practised it in the wrong way by spraying chemical pesticides in the prohibited area. Many farmers who practise refugia tend to seek yields from both Bt and non-Bt cotton crops, so they spray both. As a result, the very purpose of practising refugia is defeated. Some of the farmers were also observed to be intentionally not practising refugia, since they were unable to get any yield from non-Bt cotton crops. They also feared that practising refugia increased the attacks of the insects on Bt cotton plants. As a result, they do not see the importance of practising it.

When farmers buy the Bt cotton seed packets, the packets usually contain 450 g of Bt cotton seeds and a small packet of 120 g of non-Bt cotton seeds. The cost of a Bt cotton seed packet ranges from Rs 930 to Rs 1,000. The 120 g of the non-Bt cotton seed packet comes free of cost along with the Bt cotton seed packet. During the field survey it was observed that many farmers throw away the 120 g of non-Bt cotton seed packet that they get while purchasing the Bt cotton seed packet.

Descriptions about the practice of refugia can become clearer from Tables 6.19 and 6.20.

The farmers were asked whether they receive any guidelines on practising refugia from the seed shopkeepers. Their responses have been presented in the form of 'yes' and 'no'. It can be observed in Table 6.19 that most of the farmers responded that they do not receive any such

Table 6.19 Awareness regarding guidelines to be followed by farmers practicing cotton cultivation

Do you receive guidelines from the shopkeeper while buying Bt cotton seeds?

State	Farmer Type	Yes	No	Total
Telangana	Large	0	0	0
	Medium	1	0	1
	Semi-medium	6	27	33
	Small	8	28	36
	Marginal	5	25	30
	Total	20	80	100
Maharashtra	Large	1	1	2
	Medium	4	14	18
	Semi-medium	7	35	42
	Small	7	26	33
	Marginal	2	3	5
	Total	21	79	100

Source: Primary field survey

Table 6.20 Information regarding following of the guidelines by farmers

Do you follow the guidelines given by the shopkeepers?

State	Farmer Type	N.A.	Yes	No	Total
Telangana	Large	0	0	0	0
	Medium	0	1	0	1
	Semi-medium	27	6	0	33
	Small	28	8	0	36
	Marginal	25	4	1	30
	Total	80	19	1	100
Maharashtra	Large	1	1	0	2
	Medium	14	4	0	18
	Semi-medium	35	7	0	42
	Small	26	7	0	33
	Marginal	3	2	0	5
	Total	79	21	0	100

Source: Primary field survey

guidelines from the seed shopkeepers about Bt cotton cultivation. Some of the farmers also said that they know very well how to cultivate Bt cotton and do not need any suggestions regarding guidelines from the seed shopkeepers. When the seed shopkeepers were asked about the same, they said that they do try to inform farmers about the importance of practising refugia during Bt cotton cultivation, but still most of the farmers do not pay any attention to this information and either throw away or do not take the non-Bt cotton seed packet which comes free along with the Bt cotton seed packet.

The farmers who responded 'yes' in Table 6.19 were further asked whether they followed the guidelines prescribed by the seed shopkeepers. Table 6.20 shows the responses of the farmers in this regard.

The responses of the farmers have been recorded in the form of 'N.A.' (not applicable), 'yes' and 'no'. Farmers who responded 'no' in Table 6.19 have been considered not applicable for a response in Table 6.20, as they responded that they did not receive any guidelines from the seed shopkeepers.

Most of the farmers selected for the field survey had already said in Table 6.19 that they do not receive any guidelines about Bt cotton cultivation from the seed shopkeepers. Only a few farmers said they did receive certain guidelines from the seed shopkeepers, and they were further asked whether they follow those guidelines. Table 6.20 shows that all the farmers who responded that they received certain guidelines from the seed shopkeepers also said that they followed the particular guidelines given to them.

Refugia practice

Those cotton-growing farmers were further asked whether they are aware of the practice of refugia in Bt cotton cultivation. Table 6.21 shows the responses of the farmers in this regard.

Table 6.21 Awareness about the guidelines regarding the practice of refugia

Are you aware of the practise of refugia in Bt cotton plantations?

State	Farmer Type	Yes	No	Total
Telangana	Large	0	0	0
	Medium	1	0	1
	Semi-medium	28	5	33
	Small	23	13	36
	Marginal	18	12	30
	Total	70	30	100
Maharashtra	Large	2	0	2
	Medium	15	3	18
	Semi-medium	37	5	42
	Small	31	2	33
	Marginal	4	1	5
	Total	89	11	100

Source: Primary field survey

Table 6.22 Practice of refugia in the cotton fields

Do you practise refugia at your cotton field?

State	Farmer Type	N.A.	Yes	No	Total
Telangana	Large	0	0	0	0
	Medium	0	1	0	1
	Semi-medium	5	15	13	33
	Small	13	12	11	36
	Marginal	12	12	6	30
	Total	30	40	30	100
Maharashtra	Large	0	2	0	2
	Medium	3	2	13	18
	Semi-medium	5	21	16	42
	Small	2	16	15	33
	Marginal	1	2	2	5
	Total	11	43	46	100

Source: Primary field survey

Therefore, in Table 6.21, it can be observed that majority of farmers responded that they are aware of the practice of refugia.

Farmers who responded 'yes' in Table 6.21 indicating that they are aware of the practice of refugia in Bt cotton cultivation were further asked whether they practised refugia in their farms. Table 6.22 shows the responses of farmers to this question.

In Table 6.22, the responses of the farmers have been recorded in the form of N.A. (not applicable), 'yes' and 'no'. The farmers who responded in Table 6.21 that they are not aware of the practice have been considered not applicable for a response in Table 6.22. From Table 6.22, it can be observed that there were a significant number of farmers who said that they practise refugia.

After knowing the responses of the farmers on whether they practise refugia along with Bt cotton cultivation, it was important for us to enquire from the farmers who practise refugia whether their method is correct or not. Therefore, the farmers who responded 'yes' in Table 6.22 indicating that they practise refugia during Bt cotton cultivation were further asked whether they spray pesticides in the refuge area. The responses of the farmers to this question are shown in Table 6.23.

The responses of the farmers to the question on whether they spray pesticides in the refuge area have been recorded in the form of 'N.A.', 'yes' and 'no' in Table 6.23. The farmers who responded that they are not aware of the practice of refugia in Table 6.21 and the farmers who responded in Table 6.22 that though they are aware of it they do not practise it have been considered not applicable for a response in Table 6.23.

Table 6.23 Case of spraying of pesticides in the refuge area

Do you spray pesticides in the refuge area?

State	Farmer Type	N.A.	Yes	No	Total
Telangana	Large	0	0	0	0
	Medium	0	0	1	1
	Semi-medium	18	13	2	33
	Small	24	12	0	36
	Marginal	18	12	0	30
	Total	60	37	3	100
Maharashtra	Large	0	1	1	2
	Medium	16	1	1	18
	Semi-medium	21	15	6	42
	Small	17	13	3	33
	Marginal	3	2	0	5
	Total	57	32	11	100

Source: Primary field survey

Therefore, from Table 6.23, it can be observed that among farmers who practise refugia in their fields, more farmers spray in the refuge area, which means their method of practising refugia is not correct.

Reason for not practising refugia while cultivating Bt cotton

It was important to ask farmers who responded that they do not practise refugia in Table 6.22 why they do not do so. Table 6.24 shows the responses of these farmers, which have been recorded in the form of 'not applicable', 'increases overall cost' and 'don't see the importance'. All the farmers who responded in Table 6.21 that they are unaware of the practice of refugia have been considered 'not applicable' for a further response. 'Increases overall cost' was one of the three options in the questionnaire that farmers could choose. Technically, farmers are not supposed to seek any yield from the refuge area. But it has been observed during the field study that most of the farmers seek yields from both Bt and non-Bt cotton planted in the refuge area. So they also spray pesticide in the refuge area to kill the insects causing harm to cotton plants. As a result, the very purpose of practising refugia is defeated. Also, the wrong practice of refugia leads to an increase in the overall cost. As a result, many farmers gave up the practice.

The other option for the farmers to choose was 'don't see the importance'. This particular option indicates whether it is worth taking the risk of not spraying chemical pesticides in the refuge area, as non-spraying increases insect attacks on cotton plants. According to the cotton farmers, the bollworm is not only the insect that attacks cotton plants. Apart from bollworms,

Table 6.24 Reason for not practising refugia

State	Farmer Type	N.A.	Increases the Overall Cost	Don't See the Importance	Total
Telangana	Large	0	0	0	0
	Medium	1	0	0	1
	Semi-medium	20	6	7	33
	Small	25	11	0	36
	Marginal	24	6	0	30
	Total	70	23	7	100
Maharashtra	Large	2	0	0	2
	Medium	5	1	12	18
	Semi-medium	26	2	14	42
	Small	18	1	14	33
	Marginal	3	0	2	5
	Total	54	4	42	100

Source: Primary field survey

there are other insects which feed on cotton bolls. According to them, the practice of refugia, whereby non-Bt cotton plants are planted around Bt cotton plants, increases the attacks from other insects apart from bollworms. As a result, farmers do not see the importance of practising refugia.

It can be observed from Tables 6.21, 6.22, 6.23 and 6.24 that cotton-growing farmers have various reasons for not practising refugia. Even if they practise it, they do it incorrectly. They are not supposed to spray chemical pesticides in the refuge area. But they do, and this not only increases the overall cost of Bt cotton cultivation but also defeats the very purpose for which it has been recommended. It seems from these tables that most of the farmers have not understood the basic concept behind refugia and so practise it incorrectly. Some of the farmers have chosen not to implement it because they believe it increases their overall cost. But in reality, their overall cost increases because they do it incorrectly.

Health of farmers affected by the spraying of chemical pesticides

During the field survey, farmers were asked whether spraying of chemical pesticides causes any harm to their health. Biotechnology scientists have been claiming that with the coming of Bt cotton, the use of chemical pesticides has gone down. To check these facts, the cotton-growing farmers were asked whether Bt cotton seeds helped them in reducing chemical pesticide use. Tables 6.25 and 6.26 show the responses of the farmers regarding the health impact of spraying chemical pesticides and whether their use came down with the adoption of Bt cotton seeds.

Table 6.25 Health of farmers affected by chemical spraying

Does the spraying of chemical pesticides cause any harm to your health?

State	Farmer Type	N.A.	Yes	No	Total
Telangana	Large	0	0	0	0
	Medium	0	1	0	1
	Semi-medium	0	14	19	33
	Small	0	17	19	36
	Marginal	0	15	15	30
	Total	0	47	53	100
Maharashtra	Large	0	1	1	2
	Medium	0	12	6	18
	Semi-medium	0	31	11	42
	Small	1	23	9	33
	Marginal	0	3	2	5
	Total	1	70	29	100

Source: Primary field survey

Table 6.26 Has the spraying of chemical pesticides decreased with the advent of Bt cotton seeds?

State	Farmer Type	Spraying Increased	Spraying Decreased	Constant	Can't say	Total
Telangana	Large	0	0	0	0	0
	Medium	0	1	0	0	1
	Semi-medium	0	23	1	9	33
	Small	1	29	0	6	36
	Marginal	2	20	0	8	30
	Total	3	73	1	23	100
Maharashtra	Large	2	0	0	0	2
	Medium	17	1	0	0	18
	Semi-medium	40	1	1	0	42
	Small	32	1	0	0	33
	Marginal	5	0	0	0	5
	Total	96	3	1	0	100

Source: Primary field survey

In Table 6.25, the responses of the farmers have been recorded in the form of 'N.A.', 'yes' and 'no'. A farmer who does not spray chemical pesticides at all in his or her farmland has been considered not applicable for a further response.

Therefore, it can be seen in Table 6.25 that in the Warangal district of Telangana region, 47 per cent of the farmers said that their health was affected while spraying chemical pesticides in their fields. In Yavatmal district of Maharashtra, 70 per cent of the farmers responded that their health was affected while spraying chemical pesticides in their fields. The common health problems that they suffer because of inhaling the fumes of chemical sprays are headache, joint pain, itching in the hands, etc.

The cotton farmers were further asked whether the adoption of Bt cotton seeds helped them in reducing the use of chemical pesticides. Table 6.26 shows the responses of the farmers to this question.

In Table 6.26, the responses of the farmers have been recorded in the form of 'spraying increased', 'spraying decreased', 'constant' and 'can't say'.

From this table it can be observed that in the Warangal district of the Telangana region, most of the farmers needed less chemical pesticide spraying with the adoption of Bt cotton. But though with the adoption of Bt cotton the chemical pesticide sprayings came down, it did not stop completely. The Bt in Bt cotton only protects cotton bolls from bollworms and not from other insects like 'pindinalli' (green mosquitoes), and white flies. To eliminate those insects farmers still have to spray chemical pesticides. The sprayings might have been reduced for the time being with the adoption of Bt cotton, and it is possible that sprayings may have to increase in the future with the growing populations of other pests.

In case of the Yavatmal district of Maharashtra, Table 6.26 shows that with the adoption of Bt cotton by farmers, the spraying of chemical pesticides has increased.

Health of Bt cotton on animals

After Bt cotton's approval for commercial cultivation, many biotechnology scientists from both government and private institutes wanted to push other GM food crops into the Indian agriculture and market system. However, whenever the Indian regulatory bodies tried to approve GM food crops for commercial cultivation, civil society activist groups and some farmer unions like Bhartiya Kisan Sangh have protested in force and compelled the government to cancel the decision of the regulatory bodies. Their arguments have been that Bt cotton, which is the first GM cash crop, has not been safe for animals which grazed on them. To check the facts, cotton-growing farmers were asked whether they saw any health-related problems among animals feeding on cotton seed oil cakes and Bt cotton leaves. Tables 6.27 to 6.30 show the responses of the farmers in this regard.

In Table 6.27, it can be observed that the responses of the farmers have been recorded in the form of 'N.A.', 'yes' and 'no'. The farmers who did not have animals were considered not applicable for a response.

The cotton farmers who had animals were further asked whether there was any health impact on the animals which were fed cotton seed oil cake. As we know that more than 90 per cent of the seeds available in the market are Bt cotton seeds, it can be said with certainty that even the cotton seed oil cakes prepared nowadays will contain Bt cotton seeds. Therefore, for the

Table 6.27 Feeding animals with cotton seed oil cake

Have you ever fed your animals with cotton seed oil cake?

State	Farmer Type	N.A.	Yes	No	Total
Telangana	Large	0	0	0	0
	Medium	0	1	0	1
	Semi-medium	1	20	12	33
	Small	5	15	16	36
	Marginal	7	7	16	30
	Total	13	43	44	100
Maharashtra	Large	0	1	1	2
	Medium	0	12	6	18
	Semi-medium	4	24	14	42
	Small	4	19	10	33
	Marginal	0	4	1	5
	Total	8	60	32	100

Source: Primary field survey

Table 6.28 Health impact of Bt cotton on animals that were fed with cotton seed oil cake

Has there been any impact on the health of animals that were fed cotton seed oil cake?

State	Farmer Type	N.A.	Yes	No	Total
Telangana	Large	0	0	0	0
	Medium	0	1	0	1
	Semi-medium	1	20	12	33
	Small	5	15	16	36
	Marginal	7	7	16	30
	Total	13	43	44	100
Maharashtra	Large	1	0	1	2
	Medium	6	0	12	18
	Semi-medium	18	0	24	42
	Small	14	0	19	33
	Marginal	1	0	4	5
	Total	40	0	60	100

Source: Primary field survey

field survey, it was considered important to ask farmers about the health impact of animals feeding on cotton seed oil cake. Table 6.28 shows the responses of the farmers to this question.

In Table 6.28, it can be observed that the responses of the farmers have been recorded in the form of 'N.A.', 'yes' and 'no'. The farmers who did not have animals and farmers who do not feed their animals cotton seed oil cake have been considered not applicable for a response.

Therefore in both the regions of Telangana and Maharashtra, none of the cotton-growing farmers said that cotton seed oil cake caused any harm to the health of their animals. On the contrary, the cotton seed oil cake is fed to the animals to keep them healthy, as it provides essential nutrients like protein to the animals. According to Isaac (2002), GM crops do not always produce GM foods. For example, the GM foods for animals, like cotton seed oil cake, are highly processed and do not contain DNA or its protein. Such food, which has been processed through heating, fermenting, acidifying or refining, can no longer be identified as a GM food, as it does not have the GM DNA any longer or the GM DNA has been degraded. Therefore, if the GM DNA is no longer in its unique sequence to encode a particular protein, then there cannot be any risk to the animals consuming it as food. In other words, only if the GM food is eaten raw or unprocessed will there be a chance of ill health for the animals or any living beings.

Farmers were further asked whether they allow their animals to graze on harvested cotton fields. The responses of the farmers have been recorded in Table 6.29 in the form of 'N.A.', 'yes' and 'no'. The farmers who do not have animals have been considered not applicable for a response.

From Table 6.29, it can be observed that significant number of farmers in Telangana and most of the farmers in Maharashtra allow their animals to graze on Bt cotton leaves.

The farmers who responded in Table 6.29 that they allow their animals to graze on harvested cotton fields were further asked whether there has been

Table 6.29 Enquiring about the animals grazing on Bt cotton fields

Do you leave your animals to graze on harvested Bt cotton fields?

State	Farmer Type	N.A.	Yes	No	Total
Telangana	Large	0	0	0	0
	Medium	0	1	0	1
	Semi-medium	1	23	9	33
	Small	5	17	14	36
	Marginal	7	6	17	30
	Total	13	47	40	100
Maharashtra	Large	0	2	0	2
	Medium	0	16	2	18
	Semi-medium	4	38	0	42
	Small	4	28	1	33
	Marginal	0	5	0	5
	Total	8	89	3	100

Source: Primary field survey

Table 6.30 Health impact of Bt cotton on animals grazing on harvested cotton fields

Has there been any impact on the health of animals grazing on the harvested cotton fields?

State	Farmer Type	N.A.	Yes	No	Total
Telangana	Large	0	0	0	0
	Medium	0	1	0	1
	Semi-medium	10	9	14	33
	Small	19	8	9	36
	Marginal	24	1	5	30
	Total	53	19	28	100
Maharashtra	Large	0	1	1	2
	Medium	2	9	7	18
	Semi-medium	4	25	13	42
	Small	5	17	11	33
	Marginal	0	3	2	5
	Total	11	55	34	100

Source: Primary field survey

any impact on the health of animals grazing on harvested cotton fields. The responses of the farmers have been captured in Table 6.30.

From Table 6.30, it can be observed that some of the farmers in both Telangana and Maharashtra have reported that their animals fell sick after grazing in cotton fields. Most of the farmers said that their animals show symptoms of indigestion after eating Bt cotton leaves. To get more clarity on the matter, Dr Yakub Reddy, a veterinary doctor of the Hasanparthi block of the Warangal district of the Telangana region and Mr B. Karunakar, the livestock assistant of Mucherla village in Telangana, were contacted.

Mr B. Karunakar[6] said that animals grazing on Bt cotton fields suffer from infection in the lungs and indigestion. He said that during summer, when most grasses have dried up, animals have no option but to graze on cotton fields. It is during this time that they are seen suffering from diseases like indigestion and lung infections.

Dr Yakub Reddy[7] said that animals which graze on cotton plants have been suffering from pneumonia. But he said he is not sure whether the disease was caused by feeding on Bt cotton leaves or due to some other reason. He said that tissue samples of the diseased animals have been sent to the animal husbandry laboratories of Warangal and Hyderabad for tests to know the cause of the disease. To get further details about the progress of the tests being conducted in laboratories, the assistant directors of the animal husbandry department in Warangal and Hyderabad were contacted.

Dr B. Purinder,[8] who was the assistant director of animal husbandry in Warangal district of Telangana during the time of interview, said that it has been observed that the appetite of animals grazing on Bt cotton farms gradually start decreasing and the animals start becoming leaner day by day. To overcome this, veterinary doctors have been giving certain tonics in the initial stages. However, in the laboratory, the ongoing chemical testing has not found anything concrete. Studies are still being done to come up with some concrete reports.

Dr Shakeel Ahmad,[9] who was the assistant director of the animal husbandry department in Hyderabad during the time of interview, said that the department has not received any samples from the veterinary doctors in the districts and blocks of Telangana. He said that since the department laboratories have not received any samples of the diseased animals, no tests are being done. When he was asked whether any official reports saying that Bt cotton leaves are harmless to the animals consuming them, he replied that since no one has made any complaint or enquiry to the department of animal husbandry, no official report was ever made. He further said that initially in 2006, when some civil society organisations attempted to enquire about the issue, some proceedings took place in the department to prepare an official report on the safety of animals grazing on cotton plants. But after the bifurcation of the state into Telangana and Andhra Pradesh, no fresh enquiries came from any civil society organisations and the matter has been on hold since then.

Dr Bhim Singh Chavan,[10] the chief veterinary doctor of the Ghatanji block of the Yavatmal district of Maharashtra, responded to the issue that health problems have been occurring among animals grazing on Bt cotton plants, but these health problems have been occurring only in a few animals and not in masses. He further said that there has not been any deadly effect on animals because of grazing on Bt cotton plants. On being asked whether any test is underway in the animal husbandry department to know for sure that the health problems in animals have been caused by Bt cotton plants, he replied that since the effect is not deadly and these cases are not occurring en masse and there has been no demand from farmers or shepherds in this regard, as of now, according to his knowledge, no such test has been undertaken by the veterinary scientists in the animal husbandry department in Nagpur.

Main findings from the field survey of Telangana and Maharashtra

From the field survey in the Warangal and Yavatmal districts of Telangana and Maharashtra, respectively, it can be concluded that the major reason for crop failure in the region is the scarcity of rainfall and inadequate irrigation facilities. According to the farmers who were interviewed, if there is a scarcity of rainfall and inadequate irrigation facilities, any crop will fail, whether it is genetically modified or hybrid or indigenous. Plants, irrespective of their type, need water for growth. However, the farmers also reported that Bt cotton has aggravated the problem in the region. The claim of biotechnology scientists and corporate companies, which manufactured the new technology seeds, that they would help reduce chemical pesticide spraying is not correct. According to the cotton-growing farmers, the spraying of chemical pesticides and fertilizers has tremendously increased, especially in Maharashtra, with the adoption of Bt cotton seeds. The main reason for the adoption of Bt cotton seeds by farmers was that they were supposed to help reduce the spraying of chemical pesticides. Before the advent of Bt cotton, farmers were cultivating the long-staple hybrid variety of cotton called *G. hirsutum*. But this hybrid variety of cotton was attacked by several pests. Sap-sucking insects such as jassids, whiteflies, thrips and aphids, which were minor pests on the indigenous Desi cotton species, became major pests of *G. hirsutum*. The leaf-eating caterpillar *Spodoptera litura* and the three bollworm species (American bollworm: *Helicoverpa armigera*; pink bollworm: *Pectinophora gossypiella*; and spotted bollworm species complex: *Earias* spp.) also became major pests of *G. hirsutum*. Therefore, to give protection to the hybrid variety of cotton plants, spraying of insecticides like organophosphate and carbamate increased to an annual average of 47,100 metric tonnes from 1970 to 1980. Similarly, synthetic pyrethroids were introduced in 1980, primarily to control the pink bollworm and *S. litura* on cotton. The total pesticide usage in India increased from 67,200 metric tonnes per year in 1980 to 75,000 metric tonnes per year in 1990. Out of this total, cotton received 33,360 to 41,250 metric tonnes per year, i.e. 50 to 61 per cent of the total for a ten-year average on 7.5 million ha. This intensive usage led to high levels of resistance in insect pests of cotton such as white flies and bollworms to almost all the recommended insecticides. As a result, it significantly disrupted the equilibrium between cotton insect pests and among predators, parasitoids, parasites, and entomopathogens.

Before the introduction of Bt cotton seeds, insecticide pest management (IPM) for cotton was considered to restore the ecological balance and ensure long-term sustainable pest management. But despite the best efforts, like the highly effective participatory farmer field school (FFS) approach, IPM programs only partially succeeded. Pesticide expenditure in Indian cotton continued to increase between 1975 and 1990 and has continued to increase even after the implementation of IPM-FFS for cotton in the 1990s (Peshin, Kranthi, & Sharma, 2014). Table 6.31 shows insecticide expenditure as a percentage of variable costs of cotton production.

Therefore, Bt cotton seeds were projected as a technology that would help farmers reduce the use of chemical pesticides. Its adoption did help them achieve the purpose initially. But after time pests like bollworms started becoming resistant to the Bt cotton. As a result farmers had to again increase pesticide spraying until the second or third generation of Bt cotton in the form of Bollgard II or Bollgard III is available. The increase in pesticide spraying on Bt cotton further makes the cultivation expensive.

Table 6.31 Insecticide expenditure as a percentage of variable costs of cotton production

Year	Percentage
1974	2.1
1979	4.6
1984	11.9
1989	15.5
1994	13
1998	21.2
2002	42.0 to 50.0
2004	Between 32 and 36

Source: Peshin, Kranthi, and Sharma (2014)

The Bt cotton seed is also costlier than the non-Bt cotton seed varieties. As a result, the total input cost for cultivating cotton starts increasing. In addition, due to a scarcity of rainfall and inadequate irrigation facilities, the cotton crops wither and farmers are not able to get the expected returns. As a result, they suffer losses and go into debt.

Concern for ecology

Bt cotton seeds protect the cotton plants only from the bollworms. But bollworm is not the only insect that attacks cotton plants. According to the farmers in Yavatmal, apart from bollworms, the other insects that attack the cotton plants are 'safed marsi', 'turtura' and 'mawa' (local names of insects that attack cotton plants apart from bollworm). Similarly, farmers from the Warangal district of Telangana reported on 'pindinali' (green mosquitoes) and white flies attacking the cotton bolls. But Bt cotton can only protect the plants from bollworms and not the other insects. Therefore, to kill the other insects, farmers have to spray pesticides.

Environmentalists and scientists have argued that all these insects in the agricultural field maintain an ecological balance of the populations of each other. Disturbing the population of one phylum of insect might affect the populations of others, which ultimately can result in a new kind of problem. For example, in Punjab, the population of one kind of insect, the white fly, has suddenly increased and they are causing a lot of damage to cotton crops, In this case, Bt cotton has been of no use. Environmentalists have argued that the bollworm not only used to attack cotton but also checked the growth of white flies by feeding on them. But with the decrease in the population of bollworms, the population of other enemy insects of the cotton plant such as the white fly have dramatically risen, and Bt cotton cannot fight it.

Apart from that, it has also been observed that bollworms have started to become resistant to Bt cotton plants. Initially, Bt cotton plants were able to protect cotton bolls from bollworm attacks. But after some time, the bollworms gradually developed resistance to Bt cotton toxicity. As a result, Bt cotton is no longer effective against the bollworms. This has compelled the farmers to increase the spraying of chemical pesticides. Also, the use of Bt cotton technology has increased the dependency of farmers on biotechnology scientists and corporate companies to generate better varieties of Bt cotton plants.

Strategy of private companies

In the field survey it was found that everywhere in the market Bollgard II[11] was available. Bollgard I and non-Bt cotton seeds are not available at all in the two blocks or anywhere in the Warangal and Yavatmal districts of Telangana and Maharashtra, respectively. After the approval of Bollgard I in 2002 by regulatory bodies, there is no proper information about when Bollgard I became ineffective against bollworms and through what regulatory channel Bollgard II gained entry into the Indian seed market. Dr Haribabu Ejnavarzala, in his article "Obsolescence of First Generation GM Cotton Seed: Is It Planned?", has raised this serious issue and argued that it might be possible that this is one of the strategies of the private company Monsanto to keep its patent evergreen (Ejnavarzala, 2014). Therefore, it is the responsibility of the Indian regulatory system not only to check whether Bollgard I seeds have really become ineffective against the bollworms but also to locate the places where such loss of effectiveness has been reported. Thereafter through enquiry and investigation, the regulatory channel needs to take a decision not only to go for the second generation of GM crops in the form of Bollgard II but also to locate the areas where Bollgard II needs to be promoted. Therefore, with the adoption of GM technology, the role of the regulatory system increases.

Awareness of farmers on the guidelines of Bt cotton

Biotechnology scientists have recommended certain guidelines which need to be followed while practising Bt cotton cultivation. However, it has been found through the field survey that most of the cotton-growing farmers are unaware of those guidelines, and even when they know about them, they practise them incorrectly. One such case is the practice of refugia. While practising refugia around Bt cotton plants, farmers are not supposed to spray chemical pesticides in the refuge areas. But since they do not know the reason behind the practice of refugia and seek yields from both Bt cotton and non-Bt cotton, they spray pesticides in all the areas.

During the field survey, it was also observed that some farmers were well aware of the purpose of practising refugia. Yet they preferred not to do so because according to them, growing non-Bt cotton in the refuge area would increase insect attacks on cotton crops. They argued that the bollworm is not the only insect that damages cotton crops. Apart from the bollworm, there are other local insects as well which cause damage to cotton crops. According to them, Bt cotton might give protection to cotton from bollworms, but it is ineffective against the other cotton pests. The practice of refugia is basically taken up to control the development of resistance capability in bollworms against Bt cotton. But farmers ask: what about other insect attacks on cotton plants? If farmers are prevented from spraying in the refuge area, it would lead to an increase in the attacks of other insects on cotton plants. This is the reason why some farmers refuse to grow non-Bt cotton in the refuge area around Bt cotton crops.

Health-related controversy in Bt cotton cultivation

Health-related issues in Bt cotton cultivation have been analysed during the field survey in two phases. The first phase talks about farmers' health being affected because of the spraying of chemical pesticides in their fields. Some of the cotton-growing farmers during the survey reported that while spraying pesticides in their field, they tended to inhale the fumes and as a result would get headaches, joint pains and itching in the hands. According to most farmers selected for the survey, the adoption of Bt cotton has not helped in reducing the spraying of chemical pesticides. In comparison with earlier times, the use of chemical pesticides has further increased. So the claims of biotechnology scientists that Bt cotton reduces the use of chemical pesticides is not true in the Warangal district of Telangana and the Yavatmal district of Maharashtra.

In the second phase, farmers were asked about the health of animals feeding on cotton seed oil cake as well as Bt cotton plants. All the farmers responded that cotton seed oil cake has been good for the health of the animals, as they receive essential nutrients like proteins from it. However, many farmers said that the health of the animals was affected when they grazed on Bt cotton leaves. They showed symptoms of indigestion after eating Bt cotton leaves.

Here it is important to mention that Bt cotton seeds in the form of cotton seed oil cake cannot be considered as GM food for animals because it has been processed and as a result its DNA is degraded. So it cannot have the property of Bt cotton anymore. On the other hand, when the animals consumed Bt cotton leaves directly, which were raw and unprocessed, then the health of some of them was affected.

To get further clarity on the matter, a veterinary doctor at the Hasanparthi block of Warangal district in Telangana, Dr Yakub Reddy, and his livestock assistant, Mr B. Karunakar, were contacted. They said that animals grazing on Bt cotton fields have been found to be suffering from pneumonia and infections in the lungs. But they are not sure whether the animals were affected because of grazing on Bt cotton or some other reason. They said that they have sent the samples from the bodies of the diseased animals to the laboratories of the department of animal husbandry. The assistant directors of the department of animal husbandry at Warangal and Hyderabad were contacted for clarification on the matter. Dr B. Purinder, who is the assistant director of animal husbandry at Warangal, said that tests are still going on and up to now no concrete results have been found. Dr Shakeel Ahmad, who is the assistant director of animal husbandry in Hyderabad, said his department has not received any such samples for tests.

Similarly in Maharashtra, the chief veterinary doctor of Ghatanji block at Yavatmal district, Dr Bhim Singh Chavan, was consulted on 21 March 2016 at his office. He admitted that the animals eating Bt cotton plants have been suffering from health problems. But according to him, these problems are not occurring en masse and no animals have died on account of these problems. When asked whether any tests are being done in biotechnology or animal husbandry labs to check whether the health problems of the animals arose because of eating Bt cotton plants, he said that since there has not been any demand from the farmers or shepherds in this regard, according to his information, no such testing is underway.

Comparison of the responses of the Yavatmal farmers in Maharashtra with that of Warangal farmers in Telangana

Most of the responses of farmers received in Yavatmal have been similar to those received in Warangal. However, in Yavatmal, it was observed during the field survey that there has been a tremendous increase in the spraying of chemical pesticides and fertilizers after the adoption of Bt cotton by farmers. In contrast, in Warangal, most of the farmers reported a decrease in the spraying of chemical pesticides after the adoption of Bt cotton. They also responded that there has not been much increase in the spraying of chemical fertilizers. Farmers in the Warangal district responded by comparing the yield and net income from Bt cotton with that of hybrid cotton and said that their overall expenditure on cotton crops decreased following the adoption of Bt cotton seeds.

Similarly farmers in the Yavatmal district compared the yield and net income from Bt cotton with that of an indigenous cotton variety called 'Nanded' and other hybrid cotton crops. Since there has been a further increase in the spraying of chemical pesticides and fertilizers in the region even after the adoption of Bt cotton, the overall expenditure of farmers in this region has tremendously increased.

Both the Yavatmal district of Maharashtra and the Warangal district of Telangana are rain-fed areas and proper irrigation systems are not available. Therefore, farmers in both regions suffer losses because of the scarcity of rainfall and poor irrigation facilities. Despite this, farmers in the Warangal district in Telangana can be said to be better placed, as their overall expenditure on cotton cultivation decreased in comparison with the farmers in the Yavatmal district of Maharashtra.

Notes

1 Snowball sampling in sociology and statistics research is a non-probability sampling technique where existing study subjects recruit future subjects from among their acquaintances. Thus the sample size is said to grow like a rolling snowball.
2 According to the Agriculture Census 2010–11, farmers whose land size is less than 2.5 acres are considered to be marginal farmers. Farmers whose land size is between 2.5 acres and 5 acres are considered to be small farmers. Farmers whose land size is between 5 and 10 acres are considered to be semi-medium farmers. Farmers whose land size is from 10 to 25 acres are considered to be medium farmers. Finally, farmers whose land size is more than 25 acres are considered to be large farmers.
3 According to the Agriculture Census 2010–11, for a farmer to be called a medium farmer, his or her landholding should be more than 10 acres and less than 25 acres. It has been found that the average land size of the farmers in the Warangal district of Telangana was less than 10 acres. Therefore, most of the farmers here were either semi-medium or small or marginal.
4 Kharif season starts in April and lasts until October. Kharif crops are cultivated and harvested during this period. These crops depend highly on the monsoon rains. Cotton and paddy are considered to be the important kharif crops.
5 Rabi season starts in mid-November and ends in April/May. The main source of water for rabi crops is rainwater that has percolated into the ground. Rabi crops require irrigation.
6 Mr B. Karunakar was a livestock assistant during the time of the interview at the Mucherla block of the Warangal district of Telangana state. He was interviewed on 14 August 2015 at his office.
7 Dr Yakub Reddy was contacted by phone on 14 August 2015. He was the government veterinary doctor of the entire Hasanparthi block of Warangal district of Telangana state during the time of the interview.
8 Dr B. Purinder was contacted on phone on 17 August 2015. He was the assistant director of the animal husbandry department of the Warangal district of Telangana.
9 Dr Shakeel Ahmad was contacted on 19 August 2015 at his office in the animal husbandry department of Hyderabad. He was assistant director at the time of the interview.
10 Dr Bhim Singh Chavan was the chief veterinary doctor of the Ghatanji block of the Yavatmal district in Maharashtra during the time he was interviewed. He was contacted at his office on 21 March 2016.
11 Bollgard here is basically the variety of Bt cotton developed by corporate companies like Monsanto. The first Bt cotton crop was called Bollgard I. When Bollgard I became ineffective against the bollworms, Bollgard II was introduced into the market.

7　Summary and conclusion

The use of genetically modified (GM) technology in agriculture is one of the most controversial topics in India. The policy around it is still evolving. Both the proponents and opponents of the technology are striving to influence the policymaking process by forming alliances and coalitions with stakeholders having similar beliefs and perceptions. In the ongoing debate, a lot has been said and written about several concerns over the use of GM technology such as health, environment, ownership rights, intellectual property rights, beneficial and harmful consequences, the complex relationship between technology and human beings, etc. Therefore, it was interesting to analyse all these issues from the policy process perspective, as it has been observed that stakeholders have been using these issues to influence policymaking. The complications around GM technology in India have been analysed with the help of advocacy coalition framework (ACF) theory (Sabatier & Weible, 2007). The theory was developed by Sabatier and Jenkins-Smith in 1988. It basically talks about the process of forming different coalitions based on common beliefs and perspectives shared by the members of different interest groups. The ACF further talks about how those different coalitions come into conflict with each other and influence the policy-making processes of government bodies.

With the help of ACF theory, answers were sought to four major questions. They are:

1　Is there a possibility for coalitions (both pro- and anti-GM technology) to arrive at a consensus?
2　What is the reason behind the differing policy arguments for or against GM technology by various coalitions?
3　Does the farming community really recognise their economic interest while supporting or opposing GM technology, or are their beliefs influenced and shaped by certain organisations they are close to?
4　What is the social character and condition of people who are in the different policy coalitions formed to support or oppose GM technology?

To answer these questions, a qualitative as well as quantitative methodology was drawn up in order to understand the social and political complications around GM crops in India through ACF theory as the framework. Before theoretically analysing the political conflicts and influences of different newly formed coalitions, various literature related to science and agricultural biotechnology and reports produced by the parliament, Supreme Court, civil society organisations like Gene Campaign and corporate companies like Monsanto were collected to understand the complexity of the debate and to establish explicitly the views of the two sides on the matter. After collecting the multiple concerns on GM crops from secondary sources, they were contextualised through ACF to establish the process of influences in policymaking from recognised coalitions formed after the coming of GM crops in India.

Further, to answer those four pertinent questions, a field study was undertaken in three phases to interact with some of the major stakeholders in the debate. In the first phase, an open-ended questionnaire was prepared to interview scientists (including those from the fields of agriculture, environment and medicine), economists and members of various civil society groups. In the second and third phases, a closed-ended questionnaire was prepared to talk to 100 cotton-growing farmers each from Telangana and Maharashtra.

Theoretical analysis

As mentioned in the introduction, the ACF theory was adopted to analyse the conflicting arguments proposed by GM technology stakeholders. The theory can be understood at three levels: 'macro', 'micro' and 'meso'. These three levels are the foundation stones of ACF theory.

Policy subsystem and external factors (macro level)

The theory assumes that policymaking in a modern society is a highly complex process. Therefore, individuals or groups of individuals who intend to have any influence in this process must be specialised in the area. The policymaking participants include not only legislators, bureaucrats and judiciary but also researchers and journalists who are specialised in the policy area. The theory assumes that policy participants hold strong beliefs and are motivated to translate those beliefs into actual policy. As their beliefs are assumed to be strong and stable, it becomes difficult to bring any major change in the policy processes in a given system (Sabatier & Weible, 2007).

According to Sabatier and Weible (2007), the behaviour of most of the policy participants in the subsystem is mainly affected by two sets of exogenous factors, one of which is fairly stable and the other which is quite dynamic (see Figure 4.1). The relatively stable parameters include basic attributes of the problem, distribution of natural resources, basic constitutional structure and fundamental socio-cultural values and structures. These parameters rarely change within a time span of a decade or so, thus rarely providing impetus for policy change within a policy subsystem. On the other hand, the dynamic external factors include changes in socio-economic conditions, public opinion, governing coalitions and policy decisions in the other subsystems. The ACF hypothesises that a change in at least one of these dynamic factors is a necessary condition for a major policy change.

The individual model and belief systems (micro level)

At this level, ACF theory developers have shown how this theory is different from rational choice theory. In rational choice theory, self-interested actors rationally pursue relatively simple material interests. However, the ACF theory assumes that normative beliefs must be empirically ascertained, and therefore does not exclude the possibility of a priori altruistic behaviour. But it does stress the difficulty of changing normative beliefs among policy actors. According to this theory, each policy actor or participant sees the world through a set of perceptual filters composed of pre-existing beliefs that are difficult to alter. Therefore, it can be argued that actors from different coalitions are likely to perceive the same information in very different ways, leading to distrust.

The ACF borrows its key proposition from the "prospect theory" developed by Quattrone and Tversky in 1988 (Sabatier & Weible, 2007). According to them, actors value losses more than gains. Drawing inferences from here, the ACF argues that individuals remember defeats more than victories. As a result, this increases the density of ties among members within the same coalitions and exacerbates conflict across competing coalitions.

Advocacy coalitions (meso level)

According to the ACF, the behaviour and beliefs of any stakeholder are shaped by the particular society or groups that they belong to. When the stakeholders belonging to various groups and societies carrying different beliefs become policy participants, they naturally tend to influence the policymaking processes. The ACF assumes that policy participants strive to translate the elements of their beliefs into actual policy before their opponents can do the same. Therefore, in order to have any prospect of success, they seek to form allies, share resources and develop complementary strategies. The ACF has argued that policy participants would generally seek allies with people who hold similar beliefs among legislators, agency officials, interest group leaders, judges, researchers and intellectuals from multiple levels of government (Navneet, 2014).

Policy complications around GM technology

A policy is nothing but a plan of action chosen by an individual or a group of individuals forming a political party or business group or any cultural or ethnic or religious group. Policymaking is all about decision-making (Turner & Hulme, 1997). Therefore, to come to any definite decision or plan of action is not an easy task, particularly in a country like India where for each different issue so many players are involved.

Policy matters become complicated and time-consuming, especially when they deal with science-related matters. Dr Anitha Ramanna argues in her paper that "science policy is influenced by the choice of which scientific questions to research and which to ignore", and therefore there are various ways of processing the science that can lead to differences in outcomes (Ramanna, 2006). Today science, technology and business form the core of the new economy, and therefore they draw more attention in policymaking circles in India as elsewhere (Scoones, 2006).

According to Maarten Hajer, policymaking often takes place in a context where fixed political identities and stable communities are always assumed. It has been observed that in any given society, people normally live their lives individually in their own social networks and often without explicitly seeking representation in the sphere of formal politics in the location where they happen to live. But this political indifference can all of a sudden change with the coming of some new political issue or intervention (Hajer, 2003). According to Hajer, intended policy interventions can make people aware of what they feel attached to, thus influencing people's sense of collective identity, i.e. the awareness of what unites them and what separates them from others. To put it more precisely: "Policy discourse can be constitutive of political identities" (Hajer, 2003). Hajer, through his argument, advances the very important concept that "it is not political communities that seek political representation in order to influence policy-making. Here it is policy-making that provides the practices in which people start to deliberate and become politically active" (Hajer, 2003).

GM crop technology became a burning issue in India after Bt cotton was approved for commercial cultivation in 2001. This move by the Government of India (GOI) led to different political positions both within and outside the government. With the help of ACF, the political controversy behind GM crop technology can be analysed and understood. A flow diagram can make our task easier in this regard (see Figure 7.1).

The crux of Sabatier and Jenkins-Smith's ACF theory is that the formation of coalitions takes place among different interest groups, either in favour of or against a particular issue in order to affect policy formation in the future. In the earlier case, the flow diagram represents the changing situations with the introduction of new GM technology into Indian society. Before the introduction of this technology, there was an existing system in accordance with the conventional forms of practices in agriculture. Suddenly with the coming

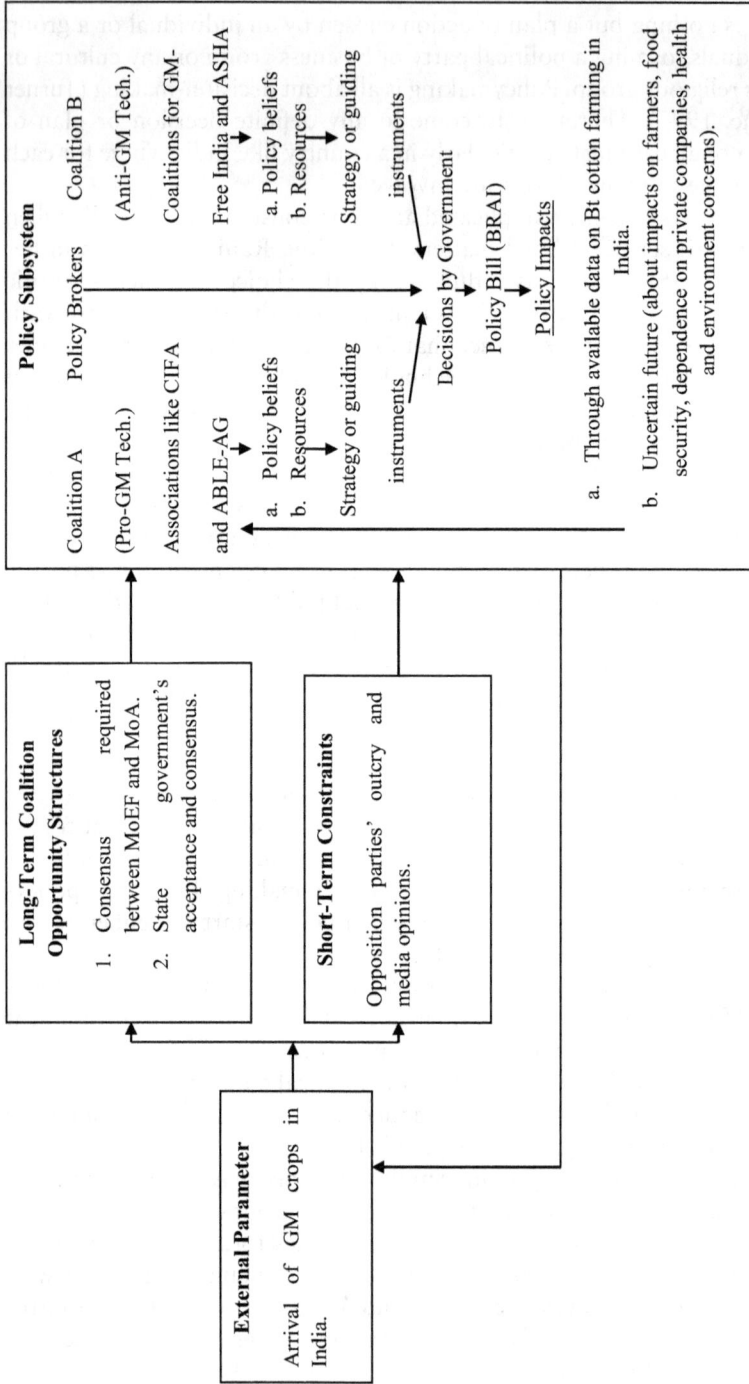

Figure 7.1 The flow diagram of the policy process of GM technology based on ACF theory

Source: Conceptual Framework developed by author based on Sabatier and Weible (2007)

of GM technology, concerns related to health, environment and ownership were raised, apart from the benefits conferred by the technology. Different concerns gave rise to a political discourse between its proponents and opponents. In the process, new political identities got recognised with the association of stakeholders with the emerging coalitions. The coalitions have been formed on the basis of similar beliefs about GM technology among different stakeholders. These stakeholders include biotechnology scientists, medical scientists, environmentalists, farmers, business groups, consumers, etc.

Keeping the ACF theory in mind, four coalitions have been identified in the Indian context which are either in support of or against the use of GM technology in agriculture. These coalitions have been categorised in Table 7.1.

The two coalitions shown in Figure 7.1 in the form of coalition A and coalition B have been further described in Table 7.1. Coalition A supports the use of GM technology in agriculture, and coalition B opposes it. The civil society groups who belong to coalition A are Shetkari Sanghatana, Punjab Agricultural University (PAU) Kisan Club, Naujawan Kisan Club, Nagarjuna Rythu Samakhya and Pratapa Rudra Farmers Mutually Aided Co-op Credit and Marketing Federation.[1] These are among the leading farmer organisations that are led by the Consortium of Farmers Association (CIFA)[2] to protest the recommendations of a Supreme Court–appointed Technical Expert Committee (TEC) for a moratorium on field trials of GM crops, as they argue that farmer communities need biotechnology to increase agricultural production.

Apart from these civil society groups, there are also some corporate-sector groups like Mahyco-Monsanto, Advanta, Kaveri Seed, Dhanuka Agritech, etc., that favour the government's decision to allow field trials of other GM seeds apart from cotton. In order to promote GM technology, biotech industries have formed the Association of Biotechnology Led Enterprises-Agriculture Group (ABLE-AG), in which biotech companies like Monsanto and Advanta are active members. ABLE projects itself as a not-for-profit pan-India forum representing the Indian biotechnology sector. It

Table 7.1 Coalitions that emerged after the adoption of GM seeds in India

Coalitions Formed to Support GM Technology	*Coalitions Formed to Oppose GM Technology*
a. Consortium of Indian Farmers Association (CIFA)	a. Coalition for a GM-Free India
b. Association of Biotech-led Enterprises – Agriculture Group (ABLE-AG)	b. Alliance for Sustainable and Holistic Agriculture (ASHA)

Source: Prepared by the author

has over 400 members from agro-biotech, bio-pharma, industrial biotech and bioinformatics sectors; investment banks and venture capital firms; leading research and academic institutes; law firms; and equipment suppliers. The primary focus of ABLE has been to accelerate the growth of the biotechnology sector in India. ABLE, in collaboration with the GoI, intends to encourage entrepreneurship and investment in the sector, provide a platform for domestic and overseas companies to explore collaboration and partnerships and forge stronger links with academia and industry and thus showcase the strengths of the Indian biotech sector. ABLE-AG consists of 12 member-companies that are considered to be the leading technology providers of modern agricultural biotechnology in the country. The members are Advanta India, BASF India, Bayer Bio Science, Devgen Seeds, Dow Agro Sciences, JK Agri Genetics, Mahyco, Metahelix, Monsanto, North Biogene, PHI Seeds and Syngenta India.

On the other side, there are some civil society groups like Gene Campaign,[3] Navdanya,[4] Greenpeace,[5] Andhra Pradesh Vyavasaya Vruthidarula Union (APVVU),[6] Shashwat Sheti Kriti Parishad (SSKP),[7] Thana (an environmental organisation in Kerala), CREATE and FEDCOT (consumer rights groups) who are protesting the move of the GoI to promote GM technology. Some non-governmental organisations (NGOs) and farmers' organisations have come together to form an association called 'Coalition for GM-Free-India'. The Coalition for GM-Free-India is a loose and informal network of various organisations and individuals from across India, campaigning and advocating for a GM-free India with the aim of shifting farming towards a sustainable path. Similarly, the Alliance for Sustainable and Holistic Agriculture (ASHA), a coalition of hundreds of organisations and individuals including numerous farmers' groups from more than 20 states, works on promoting sustainable agriculture.

Both these coalitions, which are either for or against GM crop technology, are trying to influence the Indian government in the policymaking process. The tension over the issue has been observed mainly between the Ministry of Environment Forest and Climate Change (MoEFCC) and Ministry of Agriculture (MoA). The Genetic Engineering Appraisal Committee (GEAC), which is a statutory body responsible for making the final call on whether GM crops have to be grown on Indian farms or not, falls under the ambit of MoEFCC. It has been observed that the approach of the MoEFCC office under different ministers has been widely divergent. When Jairam Ramesh and Jayanthi Natarajan were in office, they were apprehensive of giving approval to field trials for GM food crops, and a moratorium on GM crops was imposed in 2010. But later when Veerappa Moily took office during UPA's term and Prakash Javdekar during NDA's term, they removed several restrictions and approved field trials for GM crops in restricted areas like agriculture universities and research institute campuses (Navneet, 2014).

Agriculture is a subject that falls under the state list under Schedule 7 with reference to Article 246 of the Indian Constitution. Therefore, the central government alone cannot decide on the matter and state governments also need to give their consent. It has been observed that until now only four states have given no-objection certificates (NOCs) for field trials of GM crops. These states are Punjab, Delhi, Andhra Pradesh and Maharashtra. Apart from these four states, the other states are either silent on the issue or have denied permission for the field trials. The flow diagram in Figure 7.1, which is based on ACF theory, explicitly shows that the opinions of state governments as well as media and opposition would also have to be considered for transparency and accountability.

From the theoretical analysis of ACF, it appears that with the coming of GM technology in India, new political stands have been constructed to support or oppose the technology. Two coalitions, coalition A and B, supporting and opposing GM technology, respectively, have members from a wide range of sectors. Both coalitions have scientists, activists, farmers, political personalities and academicians as members to advocate for and articulate their stand on GM technology. There is no consensus on the matter within a particular professional community or group. For example, within the scientific community, there are scientists who support the use of this technology and those who oppose its use. Similarly, there are farmers' groups who want to access GM technology to increase their yield in agriculture, and there are farmers who have formed a coalition to oppose GM technology. Even within the government, there is no consensus on the matter.

The ACF theory helps explain the nature of the newly formed coalitions and the possible behaviour of the members. It also helps explain Maarten Hajer's argument that policymaking often takes place in a context where fixed political identities and stable communities are assumed, and any intervention from outside can make people aware of what they feel attached to by influencing their sense of collective identity, i.e. the awareness of what unites them and what separates them from each other (Hajer, 2003). But the theory does not dwell much on why new stands and identities are formed and what priorities the individuals or group of individuals, who recognise themselves under these new identities, have.

Millstone (2014) has talked about the priorities of different stakeholders of GM crop technology. According to him, there is sufficient evidence to show that the reason different groups of scientific assessors reach different conclusions regarding the risk from GM crops is not because "they are providing competing interpretations of agreed and shared bodies of evidence, but because they have asked and answered different questions, and have therefore reviewed different data sets; in other words they have adopted conflicting risk assessment policies" (Millstone, 2014). Millstone has elaborated on his argument by operationalising a 'co-dynamic model' (Figure 7.2).

The co-dynamic model assumes that science-based technology policymaking relies on both expert scientific assessments and non-scientific considerations. Instead of portraying risk assessment in a completely policy-free

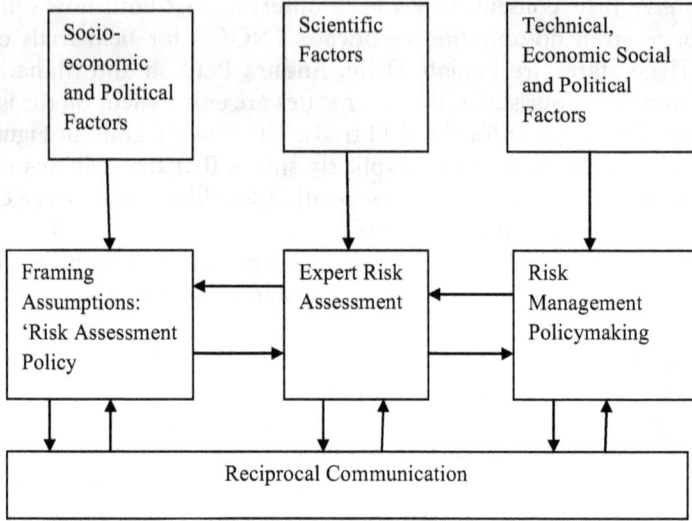

Figure 7.2 A co-dynamic model

Source: Millstone (2014)

space, the model represents the scientific deliberations as being sandwiched between two sets of judgements. The first set of judgements provides guidance as to what is to be assessed and thus seeks scientific answers to specific questions. In the second set of judgements, appropriate actions are thought of in light of scientific answers, including comparisons with other alternative courses of action, along with the acceptability of affordable costs and benefits. The different scientific voices have intensified the political conflict among coalitions of GM crop technology stakeholders.

Both the theory and the model have been used to understand the complexity of the GM debate in India. Both have been used simultaneously because it is believed that they are incomplete without each other, and the explanation of one helps in understanding the other better. The theory helps in understanding the nature of the coalition that has been built to either support or oppose GM technology on the basis of common beliefs and perceptions, while the model helps in contextualising different forms of scientific knowledge in order to shape various arguments to either support or oppose GM technology.

Methodology

As mentioned in the introduction, the field survey was conducted in three phases. In the first phase, an open-ended questionnaire was prepared to interview some policy elites like scientists from fields such as agriculture, environment and medicine; economists; and members of civil society groups. In the second and third phases, a closed-ended questionnaire was prepared to talk to 100 cotton-growing farmers each from Telangana and Maharashtra. Both the open-ended and closed-ended questionnaires were prepared to collect qualitative as well as quantitative information from the respondents. Let us see the findings of the survey briefly.

First phase of the field study

A total of 13 policy elites were interviewed, and among them most were scientists and directors of reputed research institutes as well as NGOs. Their opinions are divided, with some supporting the promotion of GM technology and some opposing it. The reason for conducting the interviews was to inquire into the factors leading to differences of opinion among them. The reason for conducting the field study was to understand their priorities and the basis for supporting or opposing GM technology and to consider how logical, sound and reasonable their arguments are.

While talking to them, several important issues were dealt with. Dr G.V. Ramanjaneyulu, who is an agriculture scientist and the executive director of an NGO called Centre for Sustainable Agriculture (CSA), says that germplasm is a common heritage of humanity and any interference in that would bring disaster. He argues that today GM technology is used as a tool to obtain exclusive rights by a few private companies over germplasm, which is a public property. He considers this to be a wrong approach and suggests that this will not make agriculture economically viable to farmers.

Dr Pushpa M. Bhargava, who was a nominee of the Supreme Court on the GEAC of the MoEFCC and was also the chairman of the Southern Regional Centre of the Council for Social Development, alleged that the regulatory bodies do not assess the technology critically. On the question why scientists' opinions are divided on the issue of GM technology, he said that scientists' opinions are not divided. He argued that all scientists who have no conflict of interest are concerned about the risks involved in the use of GM technology. These risks, according to him, are related to human, animal and plant health, as well as the environment and biodiversity. According to him, all those scientists who have no conflict of interest are actually against the open field trials of GM crops without a good safety assessment system in place.

Dr Sagari R. Ramdas, who is a veterinary scientist and is the founding member of an NGO called Anthara, is concerned about the health of the cattle feeding on Bt cotton. She and her team conducted a case study on sheep feeding on Bt cotton plants and found that these sheep are falling sick and also died in some instances.

Dr Suman Sahai is a scientist and founder of an NGO called Gene Campaign. She believes, and advocates through her organisation, that any technology must be strictly regulated in the best possible way. According to her, regulation is not being done properly by the relevant bodies in India.

She strongly believes that if a proper mechanism has not been developed to regulate sensitive technologies like GM crops, then one must not use them. According to her, technology that has raised safety concerns must be used carefully. She argues that if one can understand this in the case of nuclear technology, then there is no reason not to follow the same kind of safety practices in GM technology.

Dr Vandana Shiva is an environmentalist, activist and the director of an NGO called Navdanya and Research Foundation for Science, Technology and Natural Resource Policy. She argues that GM technologies are the new components of an old paradigm which came in the form of the Green Revolution to create a market for the chemical industry in India. GM technology has not helped farmers at all in stopping or reducing the spraying of chemical pesticides. Private companies are still selling their chemicals and at the same time getting control over seeds that farmers use in agriculture. According to her, the presumed benefits of this technology are the propaganda of private companies to maintain their monopoly in the market.

Dr Ashwani Mahajan, who is a spokesperson for Swadeshi Jagaran Manch (SJM), which is a part of the Rashtriya Swayamsewak Sangh (RSS), says that SJM opposes the use of GM technology in agriculture because it would lead to the monopoly of corporate companies over seeds. Farmers no longer have control over cotton seeds because they are developed by corporate companies, and the seeds manufactured by these companies are sterile. He argues that India does not need agriculture of the US type where companies are the owners of large landholdings and raise crops with the use of sophisticated technologies. In India, farmers have small landholdings, and there are many farmers who do not own any land and work on other farmers' land. According to him, adoption of GM technology will make farmers dependent on corporate companies, and eventually farmers who do not own any land may lose their employment in the case of crop failure due to external factors. He refers to the famous scientist Dr M.S. Swaminathan's statement that GM technology should be adopted in agriculture only if all other possibilities or alternatives have been ruled out. He further states that if there are already lots of possibilities to increase agriculture production, then there is no need to consider GM technology for agriculture. The technology is also a threat to the biodiversity of particular crops.

Mr Mohini Mohan Mishra, who is the spokesperson of Bhartiya Kissan Sangh (BKS) which is also a part of the RSS, argues that BKS opposes the use of GM technology in agriculture because it has neither helped increase crop production nor reduced the spraying of chemical pesticides. He attacks those scientists who have compared GM technologies with mobile phone technologies and says that mobile phone technologies are reversible, correctable and repairable. But GM technology cannot be reversed, corrected and repaired. When Bollgard I stopped working, scientists introduced Bollgard II. Now there is no way to recall Bollgard I. The process cannot be reversed, and there is no way to repair the damage brought about by Bt cotton to the biodiversity of cotton. So according to him, it would be foolish to compare GM technology with mobile or other technologies.

Apart from these scientists and activists, there are also scientists who support the use of GM technology and consider that it has great potential to enhance agricultural production. Dr Virander S. Chauhan, who is a scientist at the International Centre for Genetic Engineering and Biotechnology (ICGEB), says that India has a very strong biotechnology regulatory framework proposed by the department of biotechnology. But still this has not been taken up as a national policy. On the question of whether any independent research has been done to find out whether GM crops are safe for human consumption, he says that the answer to this question would come only through experiments, and therefore field trials are necessary for GM crops. He cites the example of GM corn and GM soya, which have been cultivated and consumed in the United States for 20 years. If they had been hazardous to health, by now huge amounts of data would have become available in this regard. The protests have come only from Europe, and they are unaware of the benefits of GM technology because they have not allowed experiments in the form of field trials on their land. Besides that, Dr Chauhan also mentions that many people in India do not understand science and they think GMO is only about Bt toxin. But GMO is basically about gene traits. It could have the potential to enable rice crops to become drought-resistant by placing a suitable gene into it.

Dr Ramesh Venkata Sonti is an Indian plant geneticist working as a senior scientist at the Centre for Cellular and Molecular Biology (CCMB) in Hyderabad. He was also a member of the GEAC body during the time of the interview. He argues that GM technology is not the panacea for all agricultural problems, but it does offer a solution to certain problems which may not be available through other methods.

Dr Akhilesh Kumar Tyagi, who is a scientist and was director of the National Institute of Plant Genome Research (NIPGR) at the time of the interview, thinks that scientists' opinions on the matter of GM are divided because their approaches are different, and as such they do not form one body. He feels that field trials are required by the regulatory system, as that is the only way to realise the potential of GM technology. On the question of whether any independent research has been conducted to assess the impact of GM food on animal and human health, he argues that the very definition of independence is blurred. For example, a scientist may conduct some test on biosafety which has been outsourced. So the people to whom the testing has been outsourced could be considered independent, but at the same time the other person may say that the scientists are paid.

On a similar note, Dr Ranjini Warrier, who was the director of the GEAC under the MoEFCC at the time of interview, argues that scientists are associated with some school of thought or the other in the area where they are doing research. Therefore, it is hard to ascribe to them the term 'independence' as it is normally understood. According to her, there are certain self-proclaimed independent scientists who have never visited any field and never done any plant breeding, but oppose GM technology without any basis.

Dr K.C. Bansal, who is a scientist and was the director of the National Bureau of Plant Genetic Resources (NBPGR) at the time of interview, believes that GM technology is useful for farmers and as long as this technology provides benefits, there should not be any controversy. The arguments should be based on science and facts. On the question of whether the biosafety and Cartagena protocols are taken into consideration by the regulatory bodies while assessing GM technology, he says that in the regulatory bodies all national and international protocols are considered. He believes that GM technology has the potential to enable crops to become drought-resistant, and in this way it has the potential to produce more crops per drop.

Dr Ramesh Chand, who has been the member of the National Institution for Transforming India (NITI Aayog), says that the policy around GM technology is still evolving and is not a settled issue yet. In some cases, with the use of Bt cotton seeds, the benefits were so impressive and apparent that it was accepted by the government. The government did not do anything to stop it because the benefits were clearly visible. According to him, farmers had to spray too many chemical pesticides earlier for cotton cultivation. But with the adoption of Bt cotton, there has been a tremendous decrease in the spraying of chemical pesticides, reducing the risk to farmers' health from the inhalation of chemical fumes while spraying. According to him, if the benefits generated by any technology are so great, then the government cannot stop it and has to fall in line with it in terms of policy building.

In the first phase of the field survey, it was observed that most of the government institute scientists and academicians who were interviewed were in favour of GM technology's use in agriculture. However, the scientists involved in activism or associated with any civil society groups have been raising their voice against its use in agriculture. It is significant that most of the scientists in government institutes, though they claim to be neutral, seem to be favouring the voices of coalitions like CIFA and ABLE-AG. Similarly, scientists associated with civil society groups explicitly back the voices of Coalition for GM-Free-India and ASHA.

It has been observed that coalitions such as CIFA and ABLE-AG have been closely working with the government on several fronts through the Planning Commission (during the UPA government)/NITI Aayog and academic and research institutes to encourage entrepreneurship and investment in the agricultural biotechnology sector. Their members vehemently support the use of GM technology in agriculture.

But civil society groups like CSA, Gene Campaign, Navdanya and Greenpeace are active members and supporters of ASHA and Coalition for GM-Free-India and do not receive esteem from the central government, as they are critical of its approach favouring the approval of GM crops for field trials.

The large farmers associated with coalitions like CIFA cannot be considered to represent the views of the entire farming community, as most of the farmers in India are small-scale farmers. Therefore, it is difficult to say who is right and who is wrong without trying to know the views of middle, small and marginal farmers on the issue. The picture can become clearer once the realities in agriculture and the choices of the middle, small and marginal farmers are analysed.

Summary of the field study in the second and third phases

In the second and third phases, 100 cotton-growing farmers each from Maharashtra and Telangana were selected for the field survey. During the survey, it was found that most of the responses of farmers from Telangana and Maharashtra were similar. However, in Maharashtra, it was observed that there has been a tremendous increase in the spraying of chemical pesticides and fertilizers even after the adoption of Bt cotton by farmers. In contrast, in Telangana, most of the farmers reported a decrease in the spraying of chemical pesticides after the adoption of Bt cotton. Farmers of Telangana also responded that there has not been much increase in the use of chemical fertilizers after the adoption of Bt cotton when compared to the time non-Bt cotton crops were cultivated. This might be one of the major reasons why the overall expenditure on cotton crops in Telangana decreased with the adoption of Bt cotton. But the same story is not true for Maharashtra. Here it was found that there was a tremendous increase in the use of both chemical pesticides and chemical fertilizers. As a result, the overall expenditure of farmers on cotton crops increased tremendously.

Apart from that, farmers in both regions do not have many complaints against Bt cotton seeds. They only desire to increase their net income by increasing cotton production. If any technology developed by scientists could help them increase the yield, they would not hesitate to adopt that technology. The adoption of Bt cotton is the prime example in this case. But at the same time they are not aware of the various sensitive technicalities and rules to be followed while using these technologies. For example, biotechnology scientists have recommended certain guidelines to be followed while practising Bt cotton cultivation. However, through the field survey, it has been observed that most of the cotton-growing farmers are either unaware of those guidelines, or if they know about them, they practise them wrongly. One such case is the practice of refugia. While practising refugia around Bt cotton plants, farmers are not supposed to spray chemical pesticides in the refuge areas. But since they do not know the reason for the practice of refugia and seek high yields from both Bt cotton and non-Bt cotton, they spray the pesticides in all the areas.

During the field survey, it was also found that farmers have little awareness of and concern about ecology. If any technology can help them increase their income even for a short time, they will go for it. Basically, the Bt gives cotton plants protection from only one kind of insects – the bollworm. But in the field, it was found that the bollworm is not the only insect that attacks cotton crops. Apart from bollworms, there are other insects such as 'safed marsi', 'turtura', 'mawa', 'pindinali', white flies, etc. All these insects maintain an ecological balance on the populations of each other. Disturbing the population of one phylum of insect might affect the populations of others, which ultimately can result in a new kind of problem. For example, in Punjab the recent news has been that the population of white flies has suddenly increased and they are causing a lot of damage to cotton crops. In this case, Bt cotton has been of no use. Earlier bollworms not only used to attack cotton plants but also checked the growth of white flies by feeding on them. But with the decrease in the population of bollworms, the population of other enemy insects of cotton plants has dramatically risen, and Bt cotton cannot fight these pests (Varma & Bhattacharya, 2015, October 8).

Apart from that, in the field survey it was found that the indigenous varieties of cotton seeds and other hybrid cotton seeds were not available in any of the seed shops. In all the seed shops, only Bt cotton seeds are available. Among Bt cotton seeds also, only the second generation of Bt seeds, i.e. Bollgard II, are now available, while Bollgard I seeds have almost ceased to exist. This is because initially the Bollgard I variety of Bt cotton plants were able to protect cotton bolls from the attack of bollworms. But after time, the bollworms gradually developed immunity against Bt cotton toxicity. As a result, Bollgard I lost its effectiveness against the bollworms. Poor farmers do not understand all these complications in the ecological system of the environment, and to protect their crops from attacking insects and worms, they increase the spraying of chemical pesticides. The same situation will arise again for Bollgard II eventually. Private companies will then try to launch Bollgard III. In this way, the dependence of farmers on biotechnology scientists and corporate companies to generate newer and better varieties of cotton seeds increases and the rich biodiversity of the seeds gets lost.

During the field survey in Telangana and Maharashtra, it was also found that the claims of biotechnology scientists that the adoption of Bt cotton seeds would help farmers in reducing the use of chemical pesticides spraying is not correct everywhere, especially in most parts of Maharashtra. The scientists had assumed that farmers who are prone to inhaling the fumes from chemical pesticide spraying and as a result get headaches, joint pains and itching would show an improvement in their health once Bt cotton brought about a reduction in the use of pesticides. However, this has not happened. Most of the farmers in Maharashtra reported that chemical pesticide spraying, instead of reducing, has further increased. Therefore, the claims of biotechnology scientists that Bt cotton adoption helps in the reduction of chemical pesticide use is not correct in the Vidarbha region of Maharashtra. But in Telangana, farmers reported that Bt cotton adoption has helped them to reduce chemical pesticide spraying.

Farmers who had animals were asked about the health condition of animals feeding on Bt cotton leaves. Many farmers reported that the health of their animals got worse when they grazed on Bt cotton leaves. Some farmers also responded that since they know that feeding their animals with Bt cotton leaves would affect their health, they prevented them from grazing on cotton fields. But at the same time, when the animals were fed cotton seed oil cakes, they remained fit. Citing this as an example, many scientists argue that in cotton seed oil cakes, Bt cotton is present and animals remain fit after eating it. Therefore, Bt cotton plants cannot be harmful to animals, they say.

It is important to note here that Bt cotton in the form of cotton seed oil cake cannot be considered a GM food because it has been processed, and as a result its DNA has become degraded. So it cannot have the property of Bt cotton any more (Isaac, 2002). On the other hand, when the animals consumed Bt cotton leaves directly, which were raw and unprocessed, then the health of some was affected.

To get clarity on the matter, the veterinary doctors in Telangana and Maharashtra were contacted. They admitted that animals eating Bt cotton plants have been suffering from health problems like indigestion, pneumonia and infections in their lungs. But these problems are not occurring en masse. Only a few animals have been found to show an allergic reaction to Bt cotton plants. Therefore, they are not very sure whether these health problems are occurring because of grazing on Bt cotton plants or some other reason. They said that they have sent tissue samples from the diseased animals to the laboratories of animal husbandry departments and are waiting for the test results. But when the assistant directors of the animal husbandry labs were contacted, they replied that no such tests are being done since farmers or shepherds or civil society groups have not made a concerted demand in this regard.

Farmers' perspective of Bt cotton in the realm of ACF theory

ACF draws our attention to the nature of the coalition or alliances formed by individuals or groups of individuals as stakeholders. ACF mainly studies the strengths and weaknesses of any coalition. Based on the analysis of ACF theory, four coalitions were recognised. These four coalitions are CIFA and ABLE-AG supporting the use of GM technology in agriculture and coalitions such as the Coalition for GM-Free India and ASHA opposing its use. Now it is difficult to say in which of these coalitions farmers fit in as major stakeholders. All the four mentioned coalitions project themselves as representing the interests of farmers. But in reality, all these coalitions represent the interests of the farmers only partially and not fully.

From the field survey, it can be observed that the farmers are not completely against the use of GM technology in agriculture. They only desire to increase their net income by increasing their crop yield. If any technology developed by scientists can help them in doing so, they would certainly go for it. Adoption of Bt cotton is the prime example in this case. But at the same time, they would also not like to increase their dependence on corporate sectors to access those technologies. Bt cotton came into the Indian market in such a way that it has undermined indigenous as well as hybrid varieties of cotton. At the field survey site, it was found that indigenous and other hybrid varieties of cotton are not there at all in any of the seed shops. Only Bt cotton is available.

Bt cotton has not helped farmers in Yavatmal in reducing the use of chemical pesticides. Rather, the use of chemical pesticides has further increased in the region, thus leading to an increase in the input expenditure of farmers overall. Farmers have become more vulnerable in comparison with the time when they used to cultivate indigenous varieties of cotton. When cultivating the indigenous varieties, though their cotton yield used to be low, they never faced losses and did not have to go into debt, as their expenditure on crops used to be low. So with the intention of increasing yield, hybrid varieties of cotton were launched initially. The hybrid varieties were not as strong and were also not resistant to insect attacks. Therefore to kill the attacking insects, spraying of chemical pesticides was increased, leading to an increase in the input expenditure of farmers. Though hybrid varieties helped raise cotton production, there was also a simultaneous increase in the spraying of chemical pesticides and fertilizers. The hybrid varieties of cotton can give good yield only under certain conditions, these being good rainfall, irrigation facilities and regular use of chemical pesticides and fertilizers. With the switch to hybrid varieties of cotton, the dependence of farmers has increased on companies making chemical pesticides and fertilizers. Besides, until recently farmers could save some cotton seeds for cultivation in the next season. But with the coming of Bt cotton, farmers were unable to save any seeds for another season and now have to buy seeds from the market every new season. Bt cotton came into the market as a new technology with a promise to help farmers reduce the spraying of chemical pesticides. This was also the claim of several biotechnology scientists. The technology was therefore expected to reduce the expenditure on purchasing chemical pesticides.

But this claim of scientists was not found to be correct in the case of farmers in the Vidarbha region of the Yavatmal district of Maharashtra. So the farmers are in a great dilemma: whether to support or oppose the use of GM technology in agriculture. On the one hand, they want to increase their income by increasing their yield. On the other hand, their vulnerability has also increased with their increasing dependence on corporations.

The problem is very complex, and reliance on the free market is not going to help resolve it. Therefore, the government's intervention is urgently needed to enforce regulation through its regulatory bodies.

Politics of seed companies

Peter Newell (2003) and Ian Scoones (2006) have written extensively on corporate strategies to promote GM crops in the markets of a developing country like India. Newell (2003) writes that the period from 1987 to 1989 laid the foundation for the development of a strong private-sector seed industry in India. The economic liberalisation launched in 1991 served to hasten this process. By relaxing restrictions on the activities of foreign firms and multinational companies and by giving automatic approval to foreign technology agreements, the Indian government only facilitated private-sector plant breeding. As a result, private-sector investment in the seed sector in India more than tripled between 1993 and 1997 and reached a high of Rs 19,850 million (Newell, 2003). Manju Sharma, the then DBT secretary, claimed that the new seed policy would directly promote the seed industry and suggested that the sector is now poised for a quantum jump due to the changes made in the regulations (Newell, 2003). Similarly, Mahyco's joint director of research and development, Dr Usha Zehr, claims that India has the potential to position itself as a key market for biotechnology products, as the European and North American markets have reached a saturation level (Newell, 2003).

Some multinational companies in India with large biotech portfolios are Monsanto, Syngenta, Du Pont and Aventis. Among these, Monsanto is the most high-profile and has the largest presence in India. It has a research centre in Bangalore at the Indian Institute of Science (IISC), which has often become a magnet for various protests against GM development (Newell, 2003, p. 4). Newell, in his article, explains at great length how astutely Monsanto managed to get approval for the commercial release of its Bt cotton. According to Newell (2003), in 1990 when Monsanto sought approval for Bt cotton release, its proposal was rejected by the regulatory bodies on the basis of the high cost of technology transfer. However in 1995, Mahyco, the long-established Indian seed company headed by Dr Barwale, was granted permission to import 100 grams of transgenic cotton seed as part of an agreement with Monsanto (Newell, 2003). Taking full advantage of this opportunity, Monsanto further consolidated its position by buying a 26 per cent stake in Mahyco in 1998, thus creating Mahyco-Monsanto Biotech India Ltd. (MMB) (Newell, 2003). This was an astute and strategic move by Monsanto, as Mahyco's director, Dr Barwale, is known to be very influential in the Indian regulatory system. He is a well-respected member of the Indian agricultural industry and has been honoured by the Indian government for his contributions to the agricultural sector. According to Newell, his connections within the government extend beyond the Department of Biotechnology (DBT) to many of the key agencies involved in biosafety regulations (Newell, 2003). Soon after the linking up of Monsanto with Mahyco, in March 2002, MMB's Bt cotton was approved for commercial release for a three-year trial period in six states (Newell, 2003). Newell further states that

Monsanto's position in India has become so strong that it now maintains its own "regulatory affairs" office in Delhi and engages in routine interactions with government officials over policy development. It has led attempts to get media publicity by funding public survey demonstrations in support of GM technology and has funded some selected rich farmers with large land-holdings to induce other farmers to favour GM crop development (Newell, 2003).

Implications of Bt cotton crops for Indian agriculture

Farmers have been cultivating Bt cotton for a decade in India. With the decade-long experience, supporters and detractors of GM crop technology have been engaged in building two different and contrasting narratives. Supporters have been showing the data on the immense increase in cotton production in the country with the adoption of Bt cotton seeds by farmers. They have argued that Bt cotton seeds have helped farmers reduce the use of chemical pesticides and thus saved them the money they used to spend on buying costly pesticides. It was not the government which approved Bt cotton initially. Farmers had started cultivating it illegally. Supporters of GM crop technology have been arguing that since the benefits from Bt cotton cultivation were so impressive and apparent, the government could not do much to stop it and was compelled to fall in line by approving its commercial cultivation in 2002. They argue that farmers have the right to access high-quality seeds and technologies for a higher yield in agriculture. According to them, the farmers themselves chose the higher-quality Bt cotton seeds to earn higher income.

However, detractors have drawn a contrasting picture on Bt cotton cultivation in India. They have argued that it has damaged the environment and the biodiversity of cotton crop plants. Today, the private companies have flooded the markets with Bt cotton seeds because of which other hybrid and indigenous varieties have become extinct. The detractors have been alleging that Bt cotton seeds have not helped farmers increase their yield; rather, with their adoption, farmers have gone into debt and are committing suicide in increasing numbers. The detractors have also alleged that animals have fallen sick and even died after feeding on Bt cotton plants. Thus, Bt cotton cultivation has increased the conflict among supporters and detractors of GM crop technology even further.

Conclusion

The field surveys in three phases have highlighted the different 'truths' about GM crops. These 'truths' relate to issues concerning yield, biodiversity, ecological disturbance, health and ownership rights of farmers over seeds. During the field survey, it was found that farmers are at the receiving end from the after-effects of GM technology. Also, farmers are unaware of the future consequences of the use of GM crops. Their goal is only short-term, i.e.

to get a higher yield and thus increase their income. But the real political conflict is happening among the experts of different coalitions or alliance groups mentioned in Table 7.1. It has been observed that the coalition members of both pro- and anti-GM crops have been projecting themselves as being on the side of farmers and working for their benefit. ACF theory and Millstone's co-dynamic model have been used simultaneously in the book to observe the broader picture of the complications involved in the use of GM technology in agriculture. While the ACF talks about the nature of various coalitions formed to support or oppose the use of GM technology in agriculture, the co-dynamic model elaborates on the reasons and normative judgements that brought the various stakeholders together to form a coalition. In other words, ACF theory and the co-dynamic model together would help in establishing a dialogue between stakeholders of different views. The stakeholders, who are members of different coalitions, are experts in specific areas like farming, environment, biotechnology, health and economics and therefore analyse GM crops from their own specific perspective. Frequently, experts from different areas might disagree on an issue like GM technology. Therefore, this increases the need for regulatory bodies to consider the voices of different stakeholders coming from different coalitions. The future of GM technology depends on finding a way in which different scientific truths can be brought together for analysis and discussion and reforming the regulatory bodies accordingly to reconcile the conflicting arguments of stakeholders. The theory and the model discussed in the book enable us to understand the debate better and should provide a perspective on policy processes on the use of GM crops in agriculture.

Notes

1 Shetkari Sanghatana is a non-political union of farmers formed with the aim of providing freedom of access to markets and technology for farmers. It was founded by Sharad Anantrao Joshi in the late 1970s.

   Similarly, PAU Kisan Club was started by Dr T.S. Sohal in 1966 in Barewal village near Punjab Agricultural University. Like Shetkari Sanghatana, PAU Club is also a non-political, non-profit organisation of progressive farmers of the state. Similarly, Naujawan Kisan Club is a non-political and non-profit organisation based in Punjab, and Karnail Singh is its president. Nagarjuna Rythu Samakhya and Pratapa Rudra Farmers Mutually Aided Cooperative Credit and Marketing Federation are the other two non-profit organisations for farmers based in Andhra Pradesh and Telangana regions, respectively.

2 CIFA was launched in March 2000 at Tirupathi by farmer leader Shri Sharadh Joshi. It is a coalition of various small organisations of farmers that work to enable Indian farmers to get access to modern technologies that can help them increase farm production and also get direct access to the market.

3 Gene Campaign is a research and advocacy organisation dedicated to the food and livelihood security of rural and Adivasi communities and the rights of farmers and local communities. It works with communities in villages as well as at policymaking levels to ensure the rights of farmers and local communities over their biodiversity and indigenous knowledge. Gene Campaign was set up in 1933 by Dr Suman Sahai and a group of scientists, environmentalists and economists.

4 Navdanya started as a programme of the Research Foundation for Science, Technology and Ecology (RFSTE), a participatory research initiative founded by Dr Vandana Shiva, to provide direction and support to environmental activism. The main aim of the Navdanya biodiversity conservation programme is to support local farmers and to rescue and conserve crops and plants that are being pushed to extinction and make them available through direct marketing.

5 Greenpeace India has been working on various issues related to the environment since 2001. It is a non-profit organisation with a presence in 40 countries across Europe, the Americas, Asia and the Pacific.

6 Andhra Pradesh Vyavasaya Vruthidarula Union (APVVU) is a federation of unions of agricultural workers, marginal farmers, fisherfolk and rural workers in Andhra Pradesh. It was established in 1991.

7 Shashwatsheti Kriti Parishad (SSKP) is a farmers' organisation promoted by the Yuva Rural Association (YRA). The organisation has been taking up local issues such as availability of water and seeds.

Annex A

Open-ended questionnaire
for civil society activists

Questions

1 What is the general profile of your organisation?
2 What are the main agendas or issues of your organisation?
3 Who are your major donors?
4 What have been the different achievements and failures of your organisation?
5 What is the perspective of the organisation regarding the use of genetically modified crops? What is the reason for the particular perspective?
6 What according to you is the cause of shifting from public-funded research to private-funded research in the case of biotechnology? Is it related to any substantial value/ethical problems? What are those problems?
7 What is the organisation's take on the regulatory bodies developed to assess GM crops in India?
8 According to your organisation, what kind of regulatory approach is needed – a process- or technology-based approach or a product- or application-based approach? Why?
9 If GM technology is to be discarded, what are the other alternatives to increase the agricultural production?
10 Can organic agriculture be sufficient enough in increasing the food production for self-sufficiency as well as for export?
11 What scientific evidence does your organisation have regarding the safety-related issue on GM technology, whether it is safe or not safe for human or animal consumption and what could be the possible environmental impact of its use?
12 Have you conducted any independent field study in this regard? If so, can you kindly share the data/information of this independent field survey?
13 Are your opinions based on the field study conducted by you or your organisation or based on others' opinions?
14 How are countries like the United States, Canada and Argentina that are considered to be the initiators in the use of GM technology coping with its effects?

Annex B

Open-ended questionnaire for scientists, agricultural economists and members of government regulatory bodies

Questions

1 How do you see GEAC as a unit of MoEF? Is GEAC an enforcement body or simply a consulting body?

2 Why are scientists' views so divided on the GM technology issue in India?

3 How has policy been shaped with regard to GM technology in India? Is there any constitutional provision for making decision for such issues?

4 What is the main purpose in carrying out GM field trials? Where and how are they carried out? What is the time duration for such trials? How often do these trials succeed into policy?

5 Is there any possibility for political consensus between MoEF, MoA and MoS&T on the issue of GM technology by taking into consideration the views of all the stakeholders involved?

6 Has there been any independent scientist who has been associated with neither the government nor the corporate sector who has conducted independent research on the safety-related issue of GM technology?

7 Do you see any vested interest in either promotion of or opposition to GM technology?

8 Before GM food crops can be marketed, it has to get clearances from six regulatory bodies. Are all of them functional?

9 What is your view on the BRAI bill? Can it be implemented in its present structure, or does it need any further reformation?

10 How do you see the Biosafety Protocol and the Cartagena Protocol being addressed by the regulatory bodies?

11 There has been an allegation that regulatory bodies have been approving GM crops blindly merely on the data provided by private companies, rather than going for separate independent field trials and safety tests. How far is this allegation true?

12 Can GM crops help farmers increase production in rain-fed areas? Or along with GM crops do we need other basic infrastructure like irrigation facilities also in place?

13 Is this technology going to help farmers reduce their input costs by reducing the use of chemical pesticides?

Annex C

Field survey questionnaire for farmers growing GM crops

State Name:
District Name:
Taluk Name:
Village Name:
Name of the Respondent:
Address of the Respondent:
Date of the Interview:
Time of Starting the Interview:

My name is Asheesh Navneet, and I have come from the Institute for Social and Economic Change (ISEC), a social science research organization in Bangalore. I am conducting a farmer survey to analyse the objective of my research "the effect of GM crops on farmers". The survey is going to be in the form of in-depth interviews. For this purpose, in a given district, I am going to interview 100 farmers. The districts have been selected on the basis of the larger area where farmers have been cultivating Bt cotton. These districts are Warangal in Telangana and Yavatmal in Maharashtra. I have taken Maharashtra and Telangana as two Indian states for my case study because these two states are among the largest producers of cotton in India.

The findings of this survey will be used for writing the research paper for my PhD work. This survey is an independent study and is not linked to any political party or government agency. Whatever information will be provided by the farmers will be kept strictly confidential. Participation in the field survey is voluntary, and it is entirely up to the wishes of farmers to either answer or not answer the question that I shall ask. I hope that farmers will participate in this field survey, as their participation is extremely important to collect the information that would serve the objective for this research. It would usually take 15 minutes of time to complete each interview.

1   For how long have you been living in this village?

    a)   From birth/entire life   b) Migrated

2   Gender: a) Male   b) Female

3   What is your marital status?

   a)   Married
   b)   Widowed
   c)   Divorced
   d)   Single
   e)   No response

4   What is your age?

   a)   Below 25
   b)   25 to 35
   c)   35 to 45
   d)   45 to 55
   e)   Above 55

5   Do you have a BPL card or do you fall in the range of below the poverty line?

   a)   Yes
   b)   No

6   What is your caste?
   a)   General
   b)   OBC
   c)   SC/ST
   d)   No response

7   To what level in school have you studied?

   a)   To class 10
   b)   To class 12
   c)   To graduation
   d)   Not passed class 10
   e)   Don't know to read and write

8   How many members are there in your family?

   Ans:_____

9   Is agriculture your main occupation?

   a)   Yes
   b)   No
   c)   Can't say

10   Apart from agriculture do you work in another sector to earn a living?

   a)   Yes
   b)   No

11   (If yes in Q10) Then what do you do apart from farming?

   Ans:_____

   _____

12 Are you the only member of your family engaged in agriculture?

   a) Yes
   b) No

13 (If no in Q12) Then apart from you, how many members of your household are involved in farming?

   Ans:_____

   _____

14 What is the main source of income in your household?

   a) Agriculture
   b) Service
   c) Business
   d) Others (specify): _____

15 Is agriculture an ancestral occupation of your family, or is it something
   that you have undertaken in the last few years?

   a) Ancestral farming
   b) Undertaken a few years ago/new farmers

16 Are you satisfied with farming practises?

   a) Yes
   b) No

17 (If yes in Q16) Could you tell me the main reason for being fine with the
   farming?

   a) Proud to be a farmer
   b) Traditional occupation
   c) Good social status
   d) Good income
   e) Agriculture sector has a good future
   f) I enjoy doing farming
   g) Any other
   h) No opinion

18 (If no in Q16) Could you tell me the main reason for being unsatisfied
   with the farming?

   a) Income is not good
   b) No future in farming
   c) Low social status
   d) I am highly educated and this work is not of my level
   e) Wish to do another job or business
   f) Wish to live in a city
   g) Had no other alternative/doing farming just to feed family
   h) Agriculture sector is highly stressful and risky

    i)   Any other (specify)

    j)   No opinion

19 Do you or your family own any land?

    a)   Yes

    b)   No

20 (If yes in Q19) How much land do you own?

    Ans:_____

21 (If yes in Q19) Out of the total land that you own, on how much land do you do farming?

    Ans:_____

22 (If yes in Q19) Have you rented out your land?

    a)   Yes

    b)   No

23 (If yes in Q22) How much land have you rented out?

    Ans:_____

24 (If yes in Q22) How much rent do you receive?

    a)   Estimation around: _____

    b)   No response

25 Have you or your family taken any land for agriculture on lease or rent?

    a)   Yes

    b)   No

26 (If yes in Q25) Then how much land have you taken on lease/rent for agriculture?

    Ans:_____

27 (If yes in Q25) How much rent do you pay for this land where you practise agriculture?

    a)   (In Rs per month) _____

    b)   No response

28 What kind of farmer do you describe yourself?

    a)   Owner farmer

    b)   Peasant farmer

    c)   Tenant farmer

    d)   Farm labour

    e)   Others (specify)

29 How many crops do you grow in a year?

Ans:_____

30 Mainly which crops do you grow in Kharif season?

 a) First crop: _____
 b) Second crop: _____
 c) Third crop: _____

31 Is cotton among the main crop that you grow?

 a) Yes
 b) No

32 Do you grow only cotton in your farm?

 a) Yes
 b) No

33 What kind of soil do you have for cotton cultivation?

 a) Black soil
 b) Red soil
 c) Others (specify)

34 (If no in Q32) What are the other crops that you grow along with cotton?

Ans:_____

_____

35 What kind of cotton seeds do you use for farming?

 a) Local/traditional seeds
 b) Hybrid seeds
 c) Genetically modified seeds (GM seeds)
 d) Others (specify): _____

36 (If GM seed, i.e. Bt cotton in Q34) Then what is the reason for using Bt cotton seeds?

 a) More production
 b) Safety from pests
 c) Helps in reducing the use of chemical pesticides
 d) Lack of availability of local seeds
 e) The use of Bt cotton seed is the only option
 f) More profitable
 g) Others (specify): _____

37 Which brand of Bt cotton seeds do you purchase? Do you have any idea which company produces them?

Ans:_____

_____

_____

38 Does Bt cotton cultivation require more usage of machineries like tractors, etc.?

a) Yes
b) No

39 Do you have experience using local or indigenous cotton seeds?

a) Yes
b) No

40 (If yes in Q38) What was your experience in using the local or indigenous seeds in comparison to Bt cotton seeds?

Ans:_____

_____

_____

41 Do you know what is Bt in Bt cotton, and are you aware of the purpose of putting Bt into the Bt cotton?

a) Yes
b) No

42 Traditionally you used to save seeds for cultivating them in the next season. Are you able to continue that practise still after the coming of Bt seeds into the market?

a) Yes
b) No

43 Is the growth of other crops affected in any way because of Bt cotton cultivation?

a) Yes
b) No

44 After harvesting the Bt cotton, does soil remain suitable to grow other crops?

a) Yes
b) No
c) Can't say

45 People take suggestions from different quarters to gather information about seeds, fertilizers, etc., in order to increase production. Generally from whom do you take suggestions to grow Bt cotton?

  a) Self
  b) Friends or relatives
  c) Agricultural experts
  d) Shop
  e) Government agency
  f) Kissan call centre
  g) By looking at other farmers' practises
  h) Others (specify): _____

46 From where do you purchase Bt cotton seeds?

  a) Purchase from government agency
  b) Purchase from any seed shop
  c) Purchase from private companies
  d) Exchange with other farmers
  e) Purchase from farmer cooperatives
  f) Others (specify): _____

47 What is the per unit cost of Bt cotton seeds in comparison to the local or indigenous varieties?

  Ans:_____

  _____

  _____

48 Has the use of Bt cotton seed helped in increasing the cotton production in your area?

  a) Yes
  b) No

49 What are the chemical pesticides you use to spray on your farms for cotton?

  a) Pesticide 1 .... Quantity .... Value
     .... (Rupees per unit)
  b) Pesticide 2 .... Quantity .... Value
     .... (Rupees per unit)
  c) Pesticide 3 .... Quantity .... Value
     .... (Rupees per unit)
  d) Pesticide 4 .... Quantity .... Value
     .... (Rupees per unit)
  e) Pesticide 5 .... Quantity .... Value
     .... (Rupees per unit)

50 Has the spraying of those chemical pesticides come down with the coming of Bt cotton seeds?

  a) Yes
  b) No

51 What are the chemical fertilizers you are using to grow Bt cotton?

  a) Fertilizer 1 . . . . Quantity . . . . Value
       . . . . (Rupees per unit)
  b) Fertilizer 2 . . . . Quantity . . . . Value
       . . . . (Rupees per unit)
  c) Fertilizer 3 . . . . Quantity . . . . Value
       . . . . (Rupees per unit)
  d) Fertilizer 4 . . . . Quantity . . . . Value
       . . . . (Rupees per unit)
  e) Fertilizer 5 . . . . Quantity . . . . Value
       . . . . (Rupees per unit)

52 What kind of chemical fertilizers were you using for non-Bt cotton?

  a) Fertilizer 1 . . . . Quantity . . . . Value
       . . . . (Rupees per unit)
  b) Fertilizer 2 . . . . Quantity . . . . Value
       . . . . (Rupees per unit)
  c) Fertilizer 3 . . . . Quantity . . . . Value
       . . . . (Rupees per unit)
  d) Fertilizer 4 . . . . Quantity . . . . Value
       . . . . (Rupees per unit)
  e) Fertilizer 5 . . . . Quantity . . . . Value
       . . . . (Rupees per unit)

53 While buying the Bt cotton seeds do you receive any guidelines regarding what measure you should take while growing Bt cotton seeds by shopkeepers or any of the government or private agents from where you purchase your cotton seeds?

  a) Yes
  b) No

54 (If yes in Q53) Do you follow those guidelines properly?

  a) Yes
  b) No

55 (If no in Q54) Why do you not follow those guidelines?

  a) Guidelines are complicated and you don't understand them
  b) If you follow the guidelines, it is going to increase your cost of production

c)   Other reasons (specify)_____

_____

56 Among those guidelines, have they told anything about the practice of "refugia" around Bt cotton plantations?

a)   Yes
b)   No
c)   Don't remember

57 If you are aware of "refugia", do you practise that along with growing Bt cotton?

a)   Yes
b)   No

58 (If no in Q57) Then what is the reason for not practicing the "refugia" while cultivating Bt cotton?

a)   Increases the overall cost of production
b)   Don't see the importance of practicing it
c)   Others (specify)

59 Are Bt cotton seeds comparatively costlier than traditional varieties and hybrid varieties of cotton seeds?

a)   Yes
b)   No

60 What is the per unit cost of Bt cotton seeds that you purchase?

Ans: . . . .

61 (If yes in Q59) Then do you take out a loan to purchase Bt cotton seeds?

a)   Yes (specify from whom)_____

_____

b)   No

62 (If yes in Q61) Then are you able to pay back the loans that you have taken out to purchase Bt cotton seeds?

a)   Yes
b)   No

63 Has Bt cotton cultivation helped in increasing your net income so that you can manage to clear your debt?

a)   Yes
b)   No

64 Have you experienced any failure while growing Bt cotton crops?

    a) Yes

    b) No

65 What could be the reason for the failure of growing Bt cotton according to your experience?

    a) Low rainfall

    b) Less irrigation facilities

    c) Counterfeit seeds

    d) Higher cost of Bt cotton seeds

    e) When too many farmers produce cotton, this reduces the price of cotton in the market

    f) No implementation of minimum support price by government

    g) Other reasons (specify)

66 What is the main reason for farmers committing suicide in your area?

    a) They are under debt

    b) Some other reasons (Specify)_____

    _____

    _____

67 Do you know any farmer who was a Bt cotton grower and committed suicide because of the loss in the cotton production?

    a) Yes

    b) No

68 According to you, what could be the reason for his committing suicide?

    _____

    _____

    _____

69 Are there any pests of cotton that cannot be controlled by Bt?

    a) Yes (specify) _____

    b) Not aware

70 Do you have irrigation facilities available where you grow Bt cotton?

    a) Yes

    b) No

    c) Dependent on monsoon

71 What kind of irrigation facilities do you have for cotton cultivation?

    Ans:_____

    _____

    _____

72 (If no in Q70) Then has Bt cotton seeds helped in increasing the cotton yield in comparison to the traditional varieties of seeds or hybrid seeds in rain-fed areas?

   a) Yes, it has increased
   b) No, it hasn't helped in increasing the yield
   c) Can't say

73 According to your experience in farming, can you say which cotton seed would show good growth in an unirrigated area?

   a) Bt cotton seed
   b) Hybrid cotton seed
   c) Traditional or indigenous variety of cotton seed

74 Before the coming of Bt cotton and after the coming of Bt cotton, what is your experience of the cotton yield?

   a) The yield has increased
   b) The yield decreased
   c) The yield has remained constant
   d) Can't say

75 Could you please tell how many choices are provided in terms of different types of cotton seeds in the local shops or market?

   a) No choices, only Bt cotton seeds are available
   b) Different choices are available
   c) Can't say

76 If different choices are available, what are those choices in terms of different types of seeds in the market?

   Ans:_____

   _____

   _____

77 How much cotton seed do you sow in a given acre of land?

   Ans:_____

   _____

   _____

78 Are you able to save the second generation of Bt cotton for cultivation in the next season, or for the next season do you need to buy the new Bt seeds from the market?

   a) Yes
   b) No

79 Has there been any difference in the spraying of chemical pesticides and insecticides when you used to cultivate the traditional variety of cotton crops and now in Bt cotton cultivation?

   a)   The spraying of chemical pesticides and insecticides has gone down
   b)   The spraying of chemical pesticides and insecticides has increased
   c)   The spraying of chemical pesticides and insecticides has remained constant
   d)   Can't say

80 Was the spraying of chemical pesticides and fertilizers causing any harm to your health?

   a)   Yes
   b)   No

81 The animals that you have, have you ever fed them with cotton seed oil cake?

   a)   Yes
   b)   No

82 (If yes in Q81) Then has there been any impact on their health because of consuming cotton seed oil cake?

   a)   Yes
   b)   No

83 Do you leave your animals to graze on Bt cotton leaves in the farms?

   a)   Yes
   b)   No

84 (If yes in Q83) Then has there been any effect on their health because of eating Bt cotton leaves?

   a)   Yes
   b)   No

85 (If yes in Q84) What kinds of symptoms were these animals showing?

   Ans:_____

   _____

   _____

86 What was the suggestion given by the veterinary doctor who examined these animals? Did they tell you not to graze your animals on Bt cotton leaves?

   Ans:_____

   _____

   _____

References

Adam, S., & Kriesi, H. (2007). The Network Approach. In P. A. Sabatier (Ed.), *Theories of the Policy Process* (pp. 129–154). Boulder, CO: Westview Press.

Anderson, K., & Jackson, L. A. (2006). Transgenic Crops, EU Precaution, and Developing Countries. *International Journal of Technology and Globalisation*, 2(1/2), 65–80.

Anuradha, R. (2005). *Regulatory and Governance Issuesrelating to Genetically Modified Crops and Food: An India Case Study*. New York: Case Study for the New York University Project on International Governance of Genetically Modified Organisms.

Brom, W. (2004). WTO, Public Reason and Food Public Reasoning in the 'Trade Conflict' on GM-Food. *Ethical Theory and Moral Practice*, 7(4), 417–431.

Chauhan, C. (2014, February 28). Veerappa Moily Clears Field Trials of GM Crops. Published in the India news section of the *Hindustan Times* newspaper. Retrieved December 25, 2016, from www.hindustantimes.com/india/veerappa-moily-clears-field-trials-of-gm-crops/story-Sgps5nLz9P6AtoFven4Y6H.html

Committee on Agriculture-59th Report. (2013–2014). *Cultivation of Genetically Modified Food Crops-Prospects and Effects*. Lok Sabha Secretariat, New Delhi.

Damodaran, A. (2005). Re-Engineering Biosafety Regulations in India: Towards a Critique of Policy, Law and Prescriptions. *Law, Environment and Development*, 1–20.

David, G. S., & Sai, Y. (2002). Bt Cotton: Farmers' Reactions. *Economic and Political Weekly*, 4601–4602.

Directorate of Economics and Statistics. (2012). *Agricultural Statistics at a Glance*. Directorate of Economics and Statistics, Department of Agriculture and Economics, Government of India, Ministry of Agriculture, New Delhi.

Directorate of Economics and Statistics. (2018). *Agricultural Statistics at a Glance*. Directorate of Economics and Statistics, Department of Agriculture and Economics, Government of India, Ministry of Agriculture, New Delhi.

Ejnavarzala, H. (2014). Obsolescence of First Generation GM Cotton Seed: Is It Planned? *Asian Biotechnology and Development Review*, 47–60.

Final Report of the Technical Expert Committee (TEC). (2013). *TEC Report*. Supreme Court of India, New Delhi. Retrieved from www.greenpeace.org/india/Global/india/report/2013/TEC-report.pdf

Friedman, M. (2015). GMOs: Capitalism's Distortion of Biological Processes. *Monthly Review*, 19–34.

Ghosh, P., & Ramaniah, T. (2002). Risk Assessment and Risk Management in Implementing the Cartagena Protocol. *Asia Regional Workshop* (pp. 102–121). IUCN-

Regional Biodiversity Programme-Asia, Department of Biotechnology, Government of India.

Gruere, G. P., Mehta-Bhatt, P., & Sengupta, D. (2008, October). Bt Cotton and Farmer Suicides in India: Reviewing the Evidence. *IFPRI Discussion Paper 00808*. Retrieved from https://www.organiccotton.org/.../56e7be58bcace8327de88e28e 311569e.pdf

Gupta, A. (2000). *Governing Biosafety in India: The Relevance of the Cartagena Protocol*. Harvard University: Belfer Center for Science & International Affairs. Retrieved from http://www.columbia.edu/cu/genie/pdf/GoverningBiosafe.pdf

Hajer, M. (2003). A Frame in the Fields: Policymaking and the Reinvention of Politics. In M. Hajer & H. Wagenaar (Eds.), *Deliberative Policy Analysis: Understanding Governance in the Network Society* (pp. 88–110). Cambridge: Cambridge University Press.

Haq, Z. (2012). Ministry Blames Bt Cotton for Farmer Suicides. *Hindustan Times*. Retrieved July 6, 2012, from www.hindustantimes.com/News-Feed/Business/ Ministry-blames-Bt-cotton-for-farmer-suicides/Article1-830798.aspx

Herring, R. J. (2006). Why Did 'Operation Cremate Monsanto' Fail? *Critical Asian Studies*, 467–493.

Herring, R. J., & Rao, N. (2012). On the 'Failure of Bt Cotton': Analysing a Decade of Experience. *Economic & Political Weekly*, XLVII(18), 45–53.

Human Development in Telangana State District Profiles. (2015). *Human Development in Telangana State: District Profiles*. Centre for Economic and Social Studies (CESS). Retrieved from http://www.esocialsciences.org/ Articles/show_Article.aspx?qs=ebKFqzOKbXo7se0+tFTcFgB/Qg9lMx8 H7EcQyowRRVbAjv2BNHkaakHzcZy9fkFL5aRxHR7oJK6+zUQPDCM4oebF 5ctvjE1dJQX9/n82h/8=

ICMR. (2004, April). Retrieved July 2, 2012, from Indian Council of Medical Research http://icmr.nic.in/reg_regimen.pdf

Ifft, J. (2001). *The Cotton Controversy*. Research Internship Papers, Centre for Civil Society.

Isaac, G. E. (2002). *Agricultural Biotechnology and Transatlantic Trade: Regulatory Barriers to GM Crops*. Oxon and New York: CABI.

Karihaloo, J. L., & Kumar, P. A. (2009). *Bt Cotton in India: A Status Report*. New Delhi: APCoAB and APAARI.

Kolady, D. E., & Herring, R. J. (2014). Regulation of Genetically Engineered Crops in India: Implications of Policy Uncertainty for Social Welfare, Competition, and Innovation. *Canadian Journal of Agricultural Economics*, 1–20.

Kumar, P. (2009, June 2). *Biopiracy, GM Seeds and Rural India*. Global Research. Retrieved May 2, 2012, from Global Research: www.globalresearch.ca/biopiracy- gm-seeds-and-rural-india/13820

Kuruganti, K. (2006, October). Biosafety and Beyond: GM Crops in India. *Economic and Political Weekly*, 4245–4247.

Kuruganti, K., & Ramanjaneyulu, G. (2008, March). *Genetic Engineering in Indian Agriculture: An Introductory Handbook*. Secunderabad: Centre for Sustainable Agriculture.

Lianchawii. (2005). Biosafety in India: Rethinking GMO Regulation. *Economic and Political Weekly*, 40(39), 4284–4289.

Millstone, E. (2014). Science and Politics in Indian GM Crop Regulation: A U-turn Down a Blind Alley. In R. Moor & M. V. Gowda (Eds.), *In India's Risks: Democratizing the Management of Threats to Environment, Health, and Values* (pp. 205–226). New Delhi: Oxford University Press.

Monsanto. (2010). *Faces of Farming*. Monsanto Company. Retrieved from https://materials.proxyvote.com/default.aspx?docHostID=73137

Navneet, A. (2014). A Co-Dynamic Model to Frame Controversies Over Genetically Modified Crops in India. *Asian Biotechnology and Development Review*, 16(3), 61–85. RIS.

Navneet, A. (2018). Policymaking in the Context of Contestations: GM Technology Debate in India. *Studies in Indian Politics*, 6(1), 117–131. Sage.

Navneet, A. (2019). Regulatory Approach Towards GM Technology in India, USA and EU: A Comparative Analysis. *Indian Journal of Public Administration*, 65(4), 869–884. Sage.

Nestmann, E., Copeland, T., & Hlywka, J. (2002). The Regulatory and Science-based Safety Evaluation of Genetically Modified Food Crops: A USA Perspective. In K. T. Atherton (Ed.), *Genetically Modified Crops: Assessing Safety* (pp. 1–44). London and New York: Taylor & Francis.

Newell, P. (2003). *Biotech Firms, Biotech Politics: Negotiating GMOs in India*. Brighton, Sussex: IDS Working Paper 201, Institute of Development Studies.

Paarlberg, R. L. (2001). *The Politics of Precaution: Genetically Modified Crops in Developing Countries*. Baltimore and London: The Johns Hopkins University Press.

Pal, S. (2010, March). Indian Cotton Production: Current Scenario. *Indian Textile Journal*, 120(6), 28.

Parsai, G. (2013, April 27). Global Scientists Back 10-year Moratorium on Field Trials of Bt Food Crops. *The Hindu*, S&T Agriculture.

Peshin, R., Kranthi, K. R., & Sharma, R. (2014). Pesticide Use and Experiences with Integrated Pest Management Programs and Bt Cotton in India. In R. Peshin & D. Pimentel (Eds.), *Integrated Pest Management* (pp. 269–306). Springer. Retrieved from https://www.researchgate.net/publication/266735939_Pesticide_Use_and_Experiences_with_Integrated_Pest_Management_Programs_and_Bt_Cotton_in_India

Pollack, M. A., & Shaffer, G. C. (2009). *When Cooperation Fails: The International Law and Politics of Genetically Modified Foods*. Oxford and New York: Oxford University Press.

Pray, C. E., Bengali, P., & Ramaswami, B. (2005). The Cost of Biosafety Regulations: The Indian Experience. *Quaterly Journal of International Agriculture*, 267–289.

Qaim, M., Subramanian, A., Naik, G., & Zilberman, D. (2006). Adoption of Cotton and Impact Variability: Insights from India. *Review of Agricultural Economics*, 28(1), 48–58.

Raman, R. (2017). The Impact of Genetically Modified (GM) Crops in Modern Agriculture: A Review. *GM Crops & Food: Biotechnology in Agriculture and the Food Chain*, 8(4), 195–208.

Ramanna, D. A. (2006). India's Policy on Genetically Modified Crops. *Asia Research Centre Working Paper 15*, 1–21. Retrieved from https://www.lse.ac.uk/asiaResearchCentre/_files/ARCWP15-Ramanna.pdf

Ronald, P. C., & Adamchak, R. W. (2008). *Tommorow's Table*. New York: Oxford University Press.

Sabatier, P. A., & Weible, C. M. (2007). The Advocacy Coalition Framework: Innovations and Clarifications. In P. A. Sabatier (Eds.), *Theories of the Policy Process* (pp. 189–220). Boulder, CO: Westview Press.

Sahai, S. (1999, January 16–29). *What Is Bt and What Is Terminator?* Economic and Political Weekly. Retrieved November 14, 2013, from JSTOR: www.jstor.org/stable/4407564

Sahai, S. (2004). The Agbiotech Task Force Report. *Current Science*, 87(4), 426–427.

Sahai, S., & Rahman, S. (2003). Performance of Bt Cotton: Data from First Commercial Crop. *Economic and Political Weekly*, 38(30), 3139–3141.

Salat, N. V., Salter, B., & Smets, G. (2010, September). *International Multi-Level Governance of GMOs: The EU, USA and Indian Cases.* London: Work Package 6, Activity 6.1, King's College.

Schauzu, M. (2011). The European Union's Regulatory Framework: Developments in Legislation, Safety Assessment, and Public Perception. In M. Baram & M. Bourrier (Eds.), *Governing Risk in GM Agriculture* (pp. 57–84). Cambridge: Cambridge University Press.

Scoones, I. (2006). *Science, Agriculture and the Politics of Policy: The Case of Biotechnology in India.* New Delhi: Orient Longman Private Limited.

Sethi, N. (2013, August 3). Jayanthi Natarajan Opposes Pawar's Views on GM Crops, Wants Field Trials Put on Hold. Published in the national news section of the *Hindu Newspaper*. Retrieved December 25, 2016, from www.thehindu.com/news/national/jayanthi-natarajan-opposes-pawars-views-on-gm-crops-wants-field-trials-put-on-hold/article4982776.ece

Sharma, A. B. (2010, April 5). *Making Sense of Development.* Retrieved May 30, 2012, from d-sector.org: www.d-sector.com/article-det.asp?id=1149

Sharma, D. C. (2010, March 6). Bt. Cotton Has Failed Admits Monsanto. *India Today*.

Shiva, V. (1992). The Seed and the Earth: Biotechnology and the Colonization of Regeneration. *Development Dialogue*. Retrieved November 14, 2013, from www.dhf.uu.se/pdffiler/9212/921-211.pdf

Shiva, V. (2012, July 5). *From Seeds of Suicide to Seeds of Hope: Why Are Indian Farmers Committing Suicide and How Can We Stop This Tragedy?* New Delhi, India. Retrieved from https://www.huffpost.com/entry/from-seeds-of-suicide-to_b_192419

Shiva, V., Emani, A., & Jafri, A. H. (1999). Globalization and Threat to Seed Security: Case of Transgenic Cotton Trials in India. *Economic & Political Weekly*, 34(10/11), 601–613.

Shukla, M., Al-Busaidi, K. T., Trivedi, M., & Tiwari, R. K. (2018). Status of Research, Regulations and Challenges for Genetically Modified Crops in India. *GM Crops & Food: Biotechnology in Agriculture and the Food Chain*, 9(4), 173–188. Taylor & Francis.

Sprink, T., Erikson, D., Schiemann, J., & Hartung, F. (2016). Regulatory Hurdles for Genome Editing: Process, – Vs. Product Based Approaches in Different Regulatory Context. *Plant Cell Reports*, 35(7), 1493–1506.

Stone, G. D. (2004). Biotechnology and the Political Ecology of Information in India. *Human Organization*, 63(2).

Stone, G. D. (2007). Agricultural Deskilling and the Spread of Genetically Modified Cotton in Warangal. *Current Anthropology*, 48(1), 71.

Tait, J. (2001). More Faust than Frankenstein: The European Debate About the Precautionary Principle and Risk Regulation for Genetically Modified Crops. *Journal of Risk Research*, 4(2), 175–189.

Talule, D. (2013). Political Economy of Agricultural Distress and Farmers Suicides in Maharashtra. *International Journal of Social Science & Interdisciplinary Research*, 2(2), 95–124.

Thomas, G., & Tavernier, J. D. (2017). Farmer-suicide in India: Debating the Role of Biotechnology. *Life Sciences, Society and Policy*, 13(8). Springer.

Toke, D. (2004). *The Politics of GM Food: A Comparative Study of the UK, USA and EU*. London and New York: Routledge.

Tomlinson, N. (2002). The Regulatory Requirements for Novel Foods: A European Perspective. In K. T. Atherton (Ed.), *Genetically Modified Crops: Assessing Safety* (pp. 45–62). London and New York: Taylor & Francis.

Turner, M., & Hulme, D. (1997). *Governance, Administration and Development: Making the State Work*. London: Macmillan Press LTD.

Tzotzos, G. T., Head, G. P., & Hull, R. (2009). *Genetically Modified Plants: Assessing Safety and Managing Risk*. Amsterdam and Boston: Academic Press/Elsevier.

Varma, S., & Bhattacharya, A. (2015, October 8). Whitefly Destroys 2/3rd of Punjab's Cotton Crop, 15 Farmers Commit Suicide. *Times of India*. Retrieved from https://timesofindia.indiatimes.com/india/Whitefly-destroys-2/3rd-of-Punjabs-cotton-crop-15-farmers-commit-suicide/articleshow/49265083.cms

Wu, F., & Butz, W. P. (2004). *The Future of Genetically Modified Crops: Lessons from the Green Revolution*. RAND Corporation. Retrieved from https://www.rand.org/content/dam/rand/pubs/monographs/2004/RAND_MG161.pdf

Index

Page numbers in *italics* indicate a figure and page numbers in **bold** indicate a table on the corresponding page.

Taylor & Francis Group
an **informa** business

Taylor & Francis eBooks

www.taylorfrancis.com

A single destination for eBooks from Taylor & Francis
with increased functionality and an improved user
experience to meet the needs of our customers.

90,000+ eBooks of award-winning academic content in
Humanities, Social Science, Science, Technology, Engineering,
and Medical written by a global network of editors and authors.

TAYLOR & FRANCIS EBOOKS OFFERS:

A streamlined
experience for
our library
customers

A single point
of discovery
for all of our
eBook content

Improved
search and
discovery of
content at both
book and
chapter level

REQUEST A FREE TRIAL
support@taylorfrancis.com

Routledge
Taylor & Francis Group

CRC Press
Taylor & Francis Group

For Product Safety Concerns and Information please contact our EU
representative GPSR@taylorandfrancis.com
Taylor & Francis Verlag GmbH, Kaufingerstraße 24, 80331 München, Germany

www.ingramcontent.com/pod-product-compliance
Lightning Source LLC
Chambersburg PA
CBHW060303220326
41598CB00027B/4219